Manifesto of Transdisciplinarity

SUNY series in Western Esoteric Traditions

David Appelbaum, *editor*

Manifesto of Transdisciplinarity

Basarab Nicolescu

Translated from the French
by Karen-Claire Voss

State University of New York Press

Published by
State University of New York Press, Albany

Printed in the United States of America

For information, address State University of New York Press,
90 State Street, Suite 700, Albany, NY 12207

Production by Judith Block
Marketing by Patrick Durocher

Library of Congress Cataloging-in-Publication Data

Nicolescu, Basarab.
Manifesto of transdisciplinarity / Basarab Nicolescu ; Karen-Claire Voss, translator.
p. cm. — (SUNY series in Western esoteric traditions)
Includes index.
ISBN 0-7914-5261-1 (alk. paper) — ISBN 0-7914-5262-X (pbk. : alk. paper)
1. Philosophy and science. 2. Science—Philosophy. I. Title. II. Series.

B67 .N55 2002
001—dc21 2001031190

10 9 8 7 6 5 4 3 2 1

Contents

1

To Avoid Any Misunderstanding

For the moment, the term *transdisciplinarity* retains a certain pristine charm, mostly because it has not yet been corrupted by time. It has begun spreading around the globe, popping up in unexpected places, giving voice to a lively new concept that is not yet widely understood. Even a few years ago, the term was virtually unknown, and it is still confused with two other relatively recent terms, *multidisciplinarity* and *interdisciplinarity.*

The term *transdisciplinarity* first appeared three decades ago almost simultaneously in the works of such varied scholars as Jean Piaget, Edgar Morin, and Erich Jantsch. It was coined to give expression to a need that was perceived—especially in the area of education—to celebrate the transgression of disciplinary boundaries, an act that far surpassed the multidisciplinary and the interdisciplinary approaches.

Today, the transdisciplinary approach is being rediscovered, unveiled, and utilized rapidly to meet the unprecedented challenges of our troubled world.

It wasn't so long ago that the "death of man" and the "end of history" were being proclaimed by scholars. The transdisciplinary approach enables us, instead, to discover not the death, but the resurrection of man as subject of his own discourse, and not the end of history, but the beginning of a new stage in human history. Transdisciplinary researchers increasingly appear like a new breed of contemporary knights-errant, utterly irrepressible rekindlers of hope.

As with any new philosophical movement, concomitant with this accelerated development of the transdisciplinary approach there is the danger of numerous perversions: commercialism, the co-opting of such research as a new tool for domination—adopting it as a fashionable slogan devoid of any content, pouring from emptiness into the void.

As someone who has contributed to the actual development of transdisciplinarity through both thought and action, through my own training as a quantum physicist enthusiastic about the potential role of science in today's culture, I feel an urgent need to bear witness to this situation.

If I have chosen the form of a manifesto, on the advice of numerous friends in France and in other countries, this does not mean that I have yielded to the absurd temptation to set forth some new Ten Commandments or to announce the discovery of some miraculous panacea for all the ills of the world. Rather, the axiomatic character of a manifesto—which allows it to embrace the extraordinary cultural, historical, religious, and political diversity of the many different peoples of the earth—permits an intuitive understanding of ideas that might otherwise be incomprehensible or inaccessible in a thousand scholarly works on the same subject. It is precisely because of their axiomatic character that two or three manifestos with global impact have already successfully passed the test of time. Since transdisciplinarity is intrinsically global in character, it too requires its own manifesto.

One last comment seems necessary. It is true that I have contributed substantially to several collaborative transdisciplinary enterprises, such as the founding of the Centre International de Recherches et Etudes Transdisciplinaires (CIRET) in Paris and the formulation of the Transdisciplinarity Charter, which was adopted at the time of the First World Congress of Transdisciplinarity, held at the Convento de Arrábida in Portugal in November 1994. However, the present manifesto is written according to the dictates of my conscience alone, and bears my name only.

I dedicate this manifesto to all the men and women who still have faith, in spite of everything and against all odds—especially all dogma and all ideology—in the quest for a tomorrow.

2

Tomorrow May Be Too Late

Two authentic revolutions have bracketed our century: the quantum revolution and the computer revolution.

The quantum revolution should have radically and definitively changed our vision of the world. And yet, since the beginning of the twentieth century nothing has happened. The massacre of humans by humans increases endlessly. The old viewpoint remains dominant. What's the cause of this blindness? What's the source of this neverending desire to make the old new again?

The irreducible novelty of the quantum vision affects only a very small elite: scientists on the cutting edge. The difficulty of transmitting a new hermetic language—mathematics—is certainly a considerable obstacle, but it is not insurmountable. Why is there this contempt for Nature, which we assume, without any real evidence, to be silent and impotent in regard to the pattern of the meaning of our life?

The computer revolution, which is taking place in front of our wondering and worried eyes, likewise should have led to a great liberation of time, which could then be dedicated to living our lives fully, not merely to surviving. It should have led to a sharing of knowledge between humans, as a prelude to a shared global richness. But, here too, nothing has happened. Merchants hasten to colonize cyberspace while innumerable prophets of gloom speak only of its imminent dangers. Why are we so inventive, in every situation, about expelling all possible and imaginable dangers, but so impoverished when it comes to proposing, to constructing, to building, to bringing forth that which is new and positive—not in some distant future, but in the present, here and now?

The contemporary growth of knowledge is without precedent in human history. We have explored otherwise unimaginable levels: from the infinitely small to the infinitely large, from the infinitely brief to the infinitely long. The sum of the knowledge about the universe and natural systems, accumulated during the twentieth century, far surpasses all that has been known during all other centuries combined. How is it that we know more about what we do, and less about who we are? How is it that the accelerating proliferation of disciplines makes the unity of knowledge more and more impossible to even imagine? How is it that as the exterior universe becomes more known, the meaning of our life and of our death declines into insignificance, even absurdity? Must atrophy of interior being be the price we pay for scientific knowledge? Must the individual and social happiness that scientism first promised us recede indefinitely, like a mirage?

One hears tell that humanity has always been in crisis but that it has always found a means to escape. Formerly, that affirmation was true. Today, it amounts to a lie, because for the first time in its history humanity has the possibility of complete self-destruction, with no possibility of survival.

This potential self-destruction of our species has a triple dimension: material, biological, and spiritual. In the Age of Reason, the irrational is more active than ever.

The nuclear arms accumulated on the surface of our planet could destroy it many times over, as if once were not enough. Lukewarm war has replaced the Cold War. Yesterday, arms were jealously guarded by a few powers; today, people are sent from one place on the planet to another with portable weapons under their arms; tomorrow, such weapons could be at the disposal of no matter what petty tyrant. Through what miracle of dialectic does one constantly contemplate war while speaking of peace? What's the source of the murderous folly of the human being? Where do we get our mysterious and immense capacity for forgetfulness? Today, millions of deaths, for nothing, take place before our blasé eyes in the name of fleeting ideologies and numberless conflicts whose obscure motivation eludes us.

For the first time in history, human beings can modify the genetic code of our species. In the absence of a new vision of the world, such a step forward amounts to potential biological self-destruction. We have not advanced at all on great metaphysical questions, yet we permit ourselves to intervene in the very depths of our biological being. In the name of what?

Without even getting up from our armchairs, we can experience a simulation of travel up to the speed permitted by Nature—the speed of light. The size of the Earth is progressively reduced to a point—the center of our consciousness. By an eccentric marriage between our own body and the computer, we can modify at will our sensations until we create a virtual reality, apparently more real than the reality accessible through our sense organs. An instrument for manipulating consciousness at the planetary level has thus been born, without our realizing it. In unclean hands, this instrument can lead to the spiritual self-destruction of our species.

To be sure, this potential triple self-destruction—material, biological, and spiritual—is the product of a blind but triumphant technoscience, obedient only to the implacable logic of utilitarianism. But how can we expect the blind to see?

Paradoxically, everything is in place for our self-destruction, but everything is also in place for positive change, just as there has been at other great turning points in history. The challenge of self-destruction has its counterpart in the hope of self-birth. The global challenge of death has its counterpart in a visionary, transpersonal, and planetary consciousness, which could be nourished by the miraculous growth of knowledge. We do not know which way the balance may swing. This is why we must act quickly—now. Because tomorrow may be too late.

3

The Grandeur and Decadence of Scientism

Since the beginning of time, the human spirit has been haunted by the idea of laws and order, which give meaning to the universe in which we live and to our own lives. The ancients therefore created the metaphysical, mythological, and metaphorical idea of cosmos. Indeed, they came up with a multidimensional Reality peopled with various entities, from men to gods, potentially passing through a whole series of intermediaries. These different entities all lived in their own worlds, ruled by their own laws, but everything was linked by common cosmic laws which generated a common cosmic order. Thus, the gods could intervene in the affairs of men, because men were in the image of the gods, and everything still had a meaning—somewhat hidden, but a meaning all the same.

Modern science was born through a violent break with this ancient vision of the world. It was founded on the idea—surprising and revolutionary for that era—of a total separation between the knowing subject and Reality, which was assumed to be completely independent from the subject who observed it. But, at the same time, modern science was given three fundamental postulates that would extend the quest for law and order on the plane of reason to an extreme degree:

1. that there existed universal laws, of a mathematical character;
2. that these laws could be discovered by scientific experiment;
3. that such experiments could be perfectly replicated.

Conforming to the first principle, Galileo thus elevated mathematics—an artificial language different from natural languages—to the rank of being a common language between God and men. The stunning success of classical physics, from Galileo, Kepler, and Newton to Einstein, has confirmed the accuracy of these three postulates. At the same time, these thinkers contributed to founding a paradigm of simplicity, which became predominant at the beginning of the nineteenth century. During the course of two centuries, classical physics succeeded in building a reassuring, optimistic vision concomitant with the idea of progress on the individual and social levels.

In keeping with the evidence furnished by the sense organs, classical physics is founded on the idea of continuity: one cannot pass from one point to another in space or time without passing through all the intermediary points. Moreover, physicists already had at their disposal a mathematical device founded on continuity: the infinitesimal calculus of Leibniz and Newton.

The idea of continuity is intimately connected to a key idea of classical physics: local causality. All physical phenomena are necessarily comprised of a continuous chain of causes and effects: each cause at a given point corresponds to a nearby effect, and each effect at a given point corresponds to a nearby cause. Thus, two points separated by a distance, perhaps of infinite length, in space and in time, are nevertheless linked by a continuous chain of causes and effects: there is no need at all for any direct action from a distance.

The more comprehensive causality of the ancients—for example, that of Aristotle—was reduced to only one of its aspects: local causality. Formal causality or final causality no longer had a place in classical physics. The cultural and social consequences of such reductionism, justified by the success of classical physics, are incalculable. Even today, most people who do not have an acute knowledge of philosophy consider the equivalence between "causality" and "local causality" as indisputable, so much so that in most cases, the adjective *local* is omitted.

The concept of determinism also had to make its triumphant entry into the history of ideas. The equations of classical physics are such that if one knows the positions and the speed of physical objects at a given moment, one is able to predict their position and their speed at any other moment in time. The laws of classical physics are determinist laws. Physical states are functions of positions and speeds; it follows that if one specifies the initial conditions (the physical state at a given moment of time), one can completely predict the physical state at any other given moment of time.

It is quite evident that the simplicity and aesthetic beauty of such concepts—continuity, local causality, determinism—whether or not they are operative in Nature, have fascinated the greatest minds of these last four centuries, our own included.

The final step was for science to subsume philosophy and ideology by proclaiming physics the queen of the sciences. More precisely, everything was reduced to physics; the biological and the physical appeared to be merely some evolutionary stages with one and the same foundation: physics. This step was facilitated by the overwhelming advances in physics. Thus, the scientistic ideology, which appeared as an ideology of the avant-garde and enjoyed an extraordinary development in the nineteenth century, was born.

But as a matter of fact, some unexpected perspectives disclosed themselves to the human spirit.

If the Universe was only a perfectly regulated, perfectly predictable machine, God could be relegated to the status of a simple hypothesis, unnecessary for explaining the functioning of the Universe. The Universe was suddenly desacralized, and its transcendence pushed back into the shadows of the irrational and of superstition. Nature offered herself as a mistress to man, in order to be penetrated to her depths, dominated, conquered. Even without succumbing to the temptation to psychoanalyze scientism, one is forced to note that the writings of nineteenth-century scientists about Nature abound in the most unbridled sexual allusions. Is it surprising that the femininity of the world has been disregarded, scoffed at, and forgotten in a civilization founded on conquest, domination, and usefulness at any price? As one perverse but inevitable effect, woman is generally condemned to play a minor role in social organization. In the scientistic euphoria of the period, it was quite natural to take for granted, as did Marx and Engels, the correspondences between economic, social, and historical laws and the laws of Nature. In the final analysis, all Marxist ideas are based on concepts originating within classical physics: continuity, local causality, determinism, objectivity.

If, like Nature, History is subjugated to objective and determinist laws, one can make a tabula rasa of the past, through social revolution or some other means. In fact, all that counts is the present, as the initial mechanistic condition. By assigning certain initial socially specific conditions, one can infallibly predict the future of humanity. It suffices for these initial conditions to be imposed in the name of goodness and truth—for example, in the name of liberty, equality, and fraternity—in order to build the ideal society. This experiment has been made on a global scale, with known results. How many millions of deaths have there been for dogmas? How much suffering in the name of the Good and the True? How is it that ideas, so noble in the beginning, can be transformed into their opposite?

On the spiritual level, the consequences of scientism have also been considerable: The only knowledge worthy of its name must therefore be scientific, objective; the only reality worthy of this name must be, of course, objective reality, ruled by objective laws. All knowledge other than scientific is thus cast into the inferno of subjectivity, tolerated at most as a meaningless embellishment or rejected with contempt as a fantasy, an illusion, a regression, or a product of the imagination. Even the word *spirituality* has become suspect and its use has been practically abandoned.

Objectivity, set up as the supreme criterion of truth, has one inevitable consequence: the transformation of the subject into an object. The death of the subject, which heralded all other deaths, is the price we pay for objective knowledge. The human being became an object—an object of the exploitation of man by man, an object of the experiments of ideologies that are proclaimed scientific, an object of scientific studies to be dissected, formalized, and manipulated. The Man–God has become a man–object, of which the only result can be self-destruction. The two world massacres of this century, not to mention the multiple local wars—those too, have produced innumerable corpses—are only the prelude to self-destruction on a global scale.

Or, perhaps, to self-birth.

For, all things considered, besides the immense hope that it has aroused, scientism has bequeathed us one persistent and deeply rooted idea: that of the existence of a single level of reality, in which the only concept of verticality is that of a person standing upright on a planet governed by the law of universal gravity.

4

Quantum Physics and Levels of Reality

By one of those strange coincidences, of which History alone holds the answer, quantum mechanics, World War I, and the Russian Revolution all arose at practically the same time: violence and massacres on the level of the visible, the quantum revolution on the level of the invisible. It was as if the death throes of the old world were accompanied by the discrete, barely perceptible appearance of the birth pangs of a new world. The dogmas and ideologies that have ravaged the twentieth century were derived from classical thought, founded on the concepts of classical physics; a new vision of the world would demolish the foundations of that old way of thinking.

Precisely at the beginning of the twentieth century, Max Planck was confronted by a problem in physics. Like all such problems, it seemed innocent enough at first. However, in the course of trying to resolve it, Planck made a discovery that, according to his own testimony, provoked a real inner turmoil. The reason for this was that he had unwittingly witnessed the entry of discontinuity into the realm of physics. According to

Planck's discovery, energy has a discrete, discontinuous structure. Planck's "quantum," from which we derive the name *quantum mechanics*, would revolutionize all of physics and would profoundly alter our vision of the world.

How can we understand real discontinuity? That is to say, how can we imagine that there is nothing between two points—not objects, nor atoms, nor molecules, nor particles, just nothing? Here our ordinary imagination experiences an intense vertigo, whereas mathematical language, which is based on another type of Imagination entirely, experiences no difficulty whatsoever. Galileo was right—mathematical language is of another nature than everyday human language.

To call continuity into question amounts to calling local causality into question, thus opening a veritable Pandora's box. The founders of quantum mechanics—Planck, Bohr, Heisenberg, Fermi, Einstein, Pauli, Dirac, Schrödinger, Born, de Broglie, and a few others, who incidentally all had solid philosophical backgrounds—were entirely conscious of the cultural and social implications of their discoveries. This is why they moved with great prudence, avoiding outspoken polemics. Yet, inasmuch as they were scientists, whatever their religious or philosophical convictions, they had to bow to experimental evidence and theoretical self-consistency.

Thus began an extraordinary modern *Mahabharata*-like drama, which would span the twentieth century until our time.

In order to clarify the methodology of transdisciplinarity, the author is obliged during the course of the next two or three chapters to explicate a few somewhat abstract conclusions from the field of quantum physics. The reader is therefore invited to pass through some of these theoretical considerations before entering into the heart of the matter.

The formalism of quantum mechanics and subsequently that of quantum physics (which took off after World War II with the construction of great particle accelerators) indeed tried to retain the concept of local causality as it appears on the macrophysical level, but from the beginning of quantum mechanics it was clear that a new type of causality was present at the quantum level, which is the level of the infinitely small and the infinitely brief. According to quantum mechanics, a physical quantity has several possible values, each of which is associated with a specific probability. However, in experimental measurement one obviously obtains a single result for the physical quantity in question. To abruptly deny the plurality of possible values for this physical "observable" quantity through the act of measurement may have seemed obscure, but it clearly indicated the existence of a new type of causality.

Seven decades after the birth of quantum mechanics, the nature of this new type of causality has been clarified thanks to a rigorous theorem—Bell's theorem—together with some extremely precise experiments. Thus, a new concept entered physics—that of nonseparability. In the macrophysical world, if two objects interacting in a given moment subsequently separate, they very clearly interact less and less. We think of two lovers compelled to be separated, one in one galaxy, the other in another galaxy. Under normal circumstances, their love would fade and eventually disappear.

In the quantum world, things happen differently. The quantum entities continue to interact no matter what their distance from one another. This appears contrary to our macrophysical laws. Interaction presupposes a connection, a signal, and according to Einstein's theory of relativity, this signal has a limited speed: the speed of light. Do quantum interactions break through this wall of light? Yes, if one insists on protecting local causality, at the cost of doing away with the theory of relativity. No, if one accepts the existence of a new type of causality—global causality, which concerns the system of all physical entities in its entirety. After all, this concept is not so surprising in everyday life. Any collective—a family, an enterprise, a nation—is always more than the simple sum of its parts. A mysterious factor of interaction which is not reducible to the properties of different individuals is always present in human communities, but is always rejected as the "demon of subjectivity." And it behooves us to recognize that we are very far indeed from human nonseparability on our little earth.

In any case, quantum nonseparability casts no doubt on causality itself but only on one of its forms: local causality. It does not cast any doubt on scientific objectivity either, but only on one of its forms: classical objectivity, which is based on the belief that there can be no connection present other than the local. The existence of nonlocal correlations enlarges the field of truth, of Reality. Quantum nonseparability tells us that in this world, at least at a certain level, there is a coherence, a unity of laws that assure the evolution of the totality of natural systems.

In turn, determinism, another pillar of classical thought, also crumbles.

Quantum entities—quantons—are very different from the objects of classical physics—corpuscles (or particles) and waves. If we want to rely on classical concepts, we must conclude that quantons are at the same time corpuscles and waves or, more precisely, they are neither corpuscles nor waves. If we want to talk about a wave, it is now a question of talking about a wave of probability, which allows us to calculate the probability of obtaining a final state from a particular initial state.

Quantons are characterized by a certain combination of their physical attributes, for example, such as their positions and their speeds. The mathematical relations leading to Heisenberg's celebrated uncertainty principle unequivocally indicate that it is impossible to localize a quanton at a specific point in space and in time. In other words, it is impossible to assign a specific trajectory to a quantum particle. The indeterminism that rules on the quantum level is a constituent, fundamental, irreducible indeterminism that signifies neither chance nor imprecision.

"Quantum randomness" is not "chance."

The word *chance* corresponds to the word *hazard* in English. In turn, the English word comes from the Arab *az-zahr* which is to say "play of the dice." It is certainly impossible to localize a quantum particle or to say which specific atom disintegrates at a precise moment, but this by no means signifies that the quantum event is an accidental event, owing to a play of the dice (played by whom?); put very simply, questions such as these have no meaning in the quantum world. They have no meaning because they presuppose that there must be a localizable trajectory, continuity, and local causality. In fact, the concept of "chance," like that of "necessity," is a classical concept. Quantum randomness is both chance and necessity, or, more precisely, neither chance nor necessity. Quantum randomness is really a constructive gamble, which has a meaning—that of the construction of our own macrophysical world. A finer material penetrates a grosser material. The two coexist; they cooperate in a unity that extends from the quanton to the cosmos.

"Indeterminism" is by no means "imprecision"; the confusion arises only if the concept of precision is implicitly connected, perhaps unconsciously, to the concepts of localizable trajectories, continuity, and local causality. Until now, the predictions of quantum mechanics have always been verified with great precision by countless experiments, but this precision pertains to the attributes proper to quantum entities, and not to those of classical objects. Moreover, even in the classical world the concept of precision has been radically called into question by chaos theory. One very small imprecision in the initial conditions leads to extremely divergent classical trajectories over the course of time. Chaos is embedded in the heart of determinism. Can planners of all sorts, builders of ideological, economic, and other systems, function in a world that is at once indeterminate and chaotic?

The major cultural impact of the quantum revolution has certainly raised questions for the contemporary philosophical dogma of the existence of a single level of Reality.

Here the meaning we give to the word *reality* is pragmatic and ontological at the same time.

By "Reality" (with a capital R) we intend first of all to designate that which resists our experiences, representations, descriptions, images, or mathematical formulations. Quantum physics caused us to discover that abstraction is not simply an intermediary between us and Nature, a tool for describing reality, but rather one of the constituent parts of Nature. In quantum physics, mathematical formulation is inseparable from experience. It resists in its own way, by its simultaneous concern for internal consistency and the need to integrate experimental data without destroying that self-consistency. Elsewhere as well, in so-called "virtual" reality or in computer-generated images, there are mathematical equations that resist: a single mathematical equation gives birth to an infinite series of images. In potentia, those images are already present in the equations or in the series of numbers. Abstraction therefore forms an integral part of Reality.

Insofar as Nature participates in the being of the world, one must give an ontological dimension to the concept of Reality. Nature is an immense, inexhaustible source of the unknown which even justifies the existence of science. Reality is not merely a social construction, the consensus of a collectivity, or some intersubjective agreement. It also has a trans-subjective dimension, because experimental data can ruin the most beautiful scientific theory. Alas, in the world of human beings, certain wrong-headed sociological, economic, or political theories continue to exist in spite of the multiple facts that contradict them.

By "level of Reality," we intend to designate an ensemble of systems that are invariant under certain laws: for example, quantum entities are subordinate to quantum laws, which depart radically from the laws of the physical world. That is to say that two levels of Reality are different if, while passing from one to the other, there is a break in the laws and a break in fundamental concepts (such as, for example, causality). No one has succeeded in finding a mathematical formalism that permits the difficult passage from one world to another. Semantic glosses, tautological definitions, or approximations are unable to replace a difficult mathematical formalism. There are even strong mathematical indications that the passage from the quantum world to the macrophysical world would never be possible. But there is nothing catastrophic about this. The discontinuity that is manifested in the quantum world is also manifested in the structure of the levels of Reality. That does not prevent the two worlds from coexisting. The proof: our own existence. Our bodies contain simultaneously a macrophysical structure and a quantum structure.

Levels of Reality are radically different from levels of organization as these have been defined in systemic approaches. Levels of organization do not presuppose a break with fundamental concepts: several levels of organization appear at one and the same level of Reality. The levels of organization correspond to different structurings of the same fundamental laws. For example, Marxist economy and classical physics belong to one and the same level of Reality.

The emergence of at least two different levels of Reality in the study of natural systems is a major event in the history of knowledge. It can lead us to reconsider our individual and social lives, to give a new interpretation to old knowledge, to explore the knowledge of ourselves in a different way, here and now.

The existence of different levels of Reality has been affirmed by different traditions and civilizations, but this affirmation was founded either on religious dogma or on the exploration of the interior universe.

In our century, in an effort to question the foundations of science, Husserl and other scholars have detected the existence of different levels of perception by the subject–observer of Reality. But these thinkers have been marginalized by academic philosophers and misunderstood by physicists, with each area being trapped in its respective specialization. In fact, these new thinkers were pioneers in the exploration of a multidimensional and multireferential reality, in which the human being is able to recover his place and his verticality.

5

A Stick Always Has Two Ends

Knowledge of the coexistence of the quantum world and the macrophysical world and the development of quantum physics have led, on the level of theory and scientific experiment, to the upheaval of what were formerly considered to be pairs of mutually exclusive contradictories (A and non-A): wave and corpuscle, continuity and discontinuity, separability and nonseparability, local causality and global causality, symmetry and a break in symmetry, reversibility and irreversibility of time, and so forth.

For example, equations of quantum physics comply with a group of symmetries, but their solutions break these symmetries. Similarly, a group of symmetries is supposed to depict the unification of all known physical interactions, but the symmetry must be broken in order to depict the differences among strong, weak, electromagnetic, and gravitational interactions.

The problem of the arrow of time has always fascinated thinkers. Our macrophysical level is characterized by time's irreversibility, or what we call the arrow of time. We go from birth to death, from youth to old age.

The inverse is not possible. The arrow of time is associated with entropy, with the growth of disorder. In contrast, the microphysical level is characterized by temporal invariance (reversibility of time). In most cases, everything passes as if a reel of film were running exactly the same images but backward. In the macrophysical world, there are some processes that violate this temporal invariance. Such exceptions are intimately connected to the birth of the universe, more precisely to the predominance of matter over antimatter. It is precisely on account of this small violation of temporal invariance that the universe is made up of matter and not antimatter.

Some notable efforts have been undertaken to introduce an arrow of time on the microphysical level as well, but for the moment, these efforts have not been successful. We have not been able to replace quantum mechanics with a more predictive theory. We must become accustomed to the paradoxical coexistence of the reversibility and the irreversibility of time as being merely one of the aspects caused by the coexistence of different levels of Reality. Since time is at the center of our terrestrial life, it is incumbent upon us to truly understand its place.

It should be noted that the time of physicists is only a gross approximation of the time of philosophers. No philosopher has ever been able to define the present moment properly. Saint Augustine wrote: "As for the present time, if it were always present, and did not pass at all, it would no longer be time, but eternity. Therefore, if time is only time because it passes, how can one say that it is; it is, only because it is at the point of being no more. Therefore, it is not true to say that this is time, because it tends toward non-being." The present time of the philosophers is a living time. It contains in itself both the past and the future, but is neither past nor future. Thought is powerless to apprehend all the richness of present time.

Physics abolishes this essential difference between the present, on the one hand, and the future, on the other, by replacing time with a banal timeline on which the points are successively and indefinitely presented as past, present, and future moments. Time thus becomes a simple parameter (in the same way as a position in space), which can be perfectly comprehended by thought and perfectly described on the mathematical level. At the macrophysical level this timeline is marked with an arrow indicating the passage of the past toward the future. This timeline, marked with an arrow, is therefore at once a simple mathematical representation and an anthropomorphic representation. The big surprise is to observe that even a mathematical (and thus rigorous) representation of time that is in accord with the information provided by our sense organs is put into doubt by the emergence of the quantum level, which is a level of Reality different from the macrophysical level. Does the time of the physicists retain some memory of the living time of the philosophers after all, thanks to the always unexpected intervention of Nature? After all, this paradoxical coexistence is not so surprising when we compare it with our own experience of life. We all feel that our own lifetime is living time, not merely the life of our times: Life, our own life, is something else other than an object that can be located in time and space. But the surprise is to find that a trace of this living time can be rediscovered in Nature. Could it be that Nature is not a lifeless book that has been put at our disposal to decipher, but a living book, which is still in the process of being written?

The intellectual scandal provoked by quantum mechanics consists in the fact that the pairs of contradictories that it generates are actually mutually contradictory when they are analyzed through the interpretive filter of classical logic. This logic is founded on three axioms:

1. The axiom of identity : A is A.
2. The axiom of noncontradiction : A is not non-A.
3. The axiom of the excluded middle : There exists no third term T ("T" from "third") which is at the same time A and non-A.

According to the hypothesis of a single level of Reality, the second and third axioms are obviously equivalent. The dogma of a single level of Reality, arbitrary like all dogma, is so embedded in our consciousness that even professional logicians forget to mention that these two axioms are in fact distinct and independent of each other.

If one nevertheless accepts this logic, which, after all, has ruled our thinking for two millennia and continues to dominate it today (particularly in political, social, and economic spheres), one immediately arrives at the conclusion that the pairs of contradictories advanced by quantum physics are mutually exclusive, because one cannot affirm the validity of a thing and its opposite at the same time: A and non-A. The perplexity created by this dilemma is quite understandable: If one is of sound mind, can one assert that night is day, black is white, man is woman, and life is death?

The problem may appear to be merely one of pure abstraction, of interest only to logicians, physicists, or philosophers. In what way is abstract logic important for our daily life?

Logic is the science that has for its object of study the norms of truth (or of "validity," if the word *truth* is too strong for today). Without norms, there is no order.

Without norms, there is no reading of the world; thus, there is no way to apply the idea to our life or our survival. It is therefore clear that a certain logic and even a certain vision of the world is hidden, often unconsciously, behind each action, whatever it is—whether it is the action of an individual, a collective, a nation, or a state. A certain logic is the implicit and hidden agenda that determines all social regulation.

Since the definitive formation of quantum mechanics around 1930, the founders of the new science have been acutely aware of the problem of formulating a new "quantum logic." Subsequent to the work of Birkhoff and von Neumann, a veritable flourishing of quantum logics was not long in coming. The aim of these new logics was to resolve the paradoxes that quantum mechanics had created and to attempt, to some extent, to arrive at a predictive power stronger than that afforded by classical logic.

By a lucky coincidence, this flourishing of quantum logics coincided with the flourishing of new mathematically rigorous formal logics which tried to enlarge the range of classical logic. This phenomenon is relatively new because for two millennia human beings had believed that logic was unique, unchangeable, given once and for all, inherent in their own minds.

However, there is a direct relation between logic and the environment—physical, chemical, biological, macro- or microsociological. Now, like knowledge and comprehension, the environment changes with time. Therefore, logic can only have an empirical foundation. The notion of the history of logic is very recent—it appeared in the middle of the nineteenth century. Another major idea—that of the history of the universe—appeared shortly afterward. Previously, the universe, like logic, had been considered to be eternal and unchangeable.

Most quantum logics have modified the second axiom of classical logic—the axiom of noncontradiction—by introducing noncontradiction with several truth values in place of the binary pair (A and non-A). These multivalent logics, whose status with respect to their predictive ability remains controversial, have not taken into account one other possibility: the modification of the third axiom—the axiom of the excluded middle.

History will credit Stéphane Lupasco with having shown that the logic of the included middle is a true logic, formalizable and formalized, multivalent (with three values: A, non-A, and T) and noncontradictory. Lupasco, like Edmund Husserl, was a pioneer. His philosophy, which takes quantum physics as its point of departure, has been marginalized by physicists and philosophers. Curiously, on the other hand, it has had a powerful, albeit underground, impact among psychologists, sociologists, artists, and historians of religion. Lupasco was right too soon. Perhaps the absence of the notion of "levels of Reality" in his philosophy obscured the substance of his philosophy. Many persons believed that Lupasco's logic violated the axiom of noncontradiction—hence the rather unfortunate name "logic of contradiction"—and that it entailed the risk of endless semantic glosses. Still more, the visceral fear of introducing the idea of the included middle, with its magical resonances, only helped to increase the distrust of such a logic.

Our understanding of the axiom of the included middle—there exists a third term T, which is at the same time A and non-A—is completely clarified once the notion of "levels of Reality" is introduced.

In order to obtain a clear image of the meaning of the included middle, we can represent three terms of the new logic—A, non-A, and T—and the dynamics associated with them by a triangle in which one of the vertices is situated at one level of Reality and the two other vertices at another level of Reality. The included middle is really an included third. If one remains at a single level of Reality, all manifestation appears as a struggle between two contradictory elements (example: wave A and corpuscle non-A). The third dynamic, that of the T-state, is exercised at another level of Reality, where that which appears to be disunited (wave or corpuscle) is in fact united (quanton), and that which appears contradictory is perceived as noncontradictory.

It is the projection of the T-state onto the same single level of Reality that produces the appearance of mutually exclusive, antagonistic pairs (A and non-A). A single level of Reality can only create antagonistic oppositions. It is inherently self-destructive if it is completely separated from all the other levels of Reality. A third term, let us call it T_0, which is situated at the same level of Reality as that of the opposites A and non-A, cannot accomplish their reconciliation. The "synthesis" between A and non-A is above all an explosion of immense energy, like that produced by the encounter between matter and antimatter. In the hands of Marxist–Leninists, the Hegelian synthesis appeared like the grand finale of progressive development on the historical plane: primitive society (thesis)—capitalist society (antithesis)—communist society (synthesis). Alas, it has metamorphosed into its opposite. The unexpected fall of the Soviet empire was in fact inexorably inscribed in the binary logic of its own system. A logic is never innocent. It can even cause millions of deaths.

The entire difference between a triad of the included middle and a Hegelian triad is clarified by consideration of the role of time. In a triad of the included middle, the three terms coexist at the same moment in time.

This is why the Hegelian triad is incapable of accomplishing the reconciliation of opposites, whereas the triad of the included middle is capable of it. In the logic of the included middle the opposites are, rather, contradictories: the tension between contradictories builds a unity that includes and goes beyond the sum of the two terms.

One also sees the great dangers of misunderstanding engendered by the common enough confusion between the axiom of the excluded middle and the axiom of noncontradiction. The logic of the included middle is noncontradictory in the sense that the axiom of noncontradiction is thoroughly respected, a condition that enlarges the notions of "true" and "false" in such a way that the rules of logical implication concerns not just two terms (A and non-A), but three terms (A, non-A, and T), all coexisting at the same moment in time. This is a formal logic, just as any other formal logic: its rules are derived by means of a relatively simple mathematical formalism.

One can see why the logic of the included middle is not simply a metaphor, like some kind of arbitrary ornament for classical logic that would permit adventurous incursions and passages into the domain of complexity. The logic of the included middle is perhaps the privileged logic of complexity; privileged in the sense that it allows us to cross the different areas of knowledge in a coherent way.

The logic of the included middle does not abolish the logic of the excluded middle: it only constrains its sphere of validity. The logic of the excluded middle is certainly valid for relatively simple situations, for example, driving a car on a highway: no one would dream of introducing an included middle in regard to what is permitted and what is prohibited in such circumstances. On the contrary, the logic of the excluded middle is harmful in complex cases, for example, within the social or political spheres. In such cases it operates like a genuine logic of exclusion: good or evil, right or left, women or men, rich or poor, whites or blacks. It would be revealing to undertake an analysis of xenophobia, racism, antisemitism, or nationalism in the light of the logic of the excluded middle. It would also be very instructive to examine the speeches of politicians through the filter of that logic.

Popular wisdom expresses something very profound when it tells us that "a stick always has two ends." Let us imagine, as in the story "Le bout du bout" (The End of the End) told by the famous French artist Raymond Devos (who seems to have understood better than have most scholars the meaning of the included middle), a man who desperately wants to separate the two ends of a stick. He cuts his stick and then sees that instead of having separated the two ends, he now has two sticks, both of which have two ends of their own. He goes on cutting his stick, all the while becoming more and more anxious—the sticks multiply ad infinitum, but he finds it impossible to separate the two ends!

Are we—in our present civilization—in the same position as this man who applies himself so assiduously to trying to separate the two ends of his stick? Only the intelligence of an included third can answer the barbarism of the excluded middle. Because a stick always has two ends.

6

The Emergence of Complex Plurality

At the same time that different levels of Reality and new logics (among them the logic of the included middle) were emerging within the study of natural systems, a third factor appeared that gave the coup de grâce to the classical vision of the world: complexity.

In the course of the twentieth century, complexity—frightful, terrifying, obscene, fascinating, and invasive—has established itself everywhere as a challenge not only to our existence itself, but also to its very meaning. Meaning seems to get absorbed as if by the white blood cell of complexity in all areas of knowledge.

This complexity is nourished by the disciplinary research boom, which, in turn, leads to the accelerating proliferation of disciplines.

Classical binary logic confers its patent on either a scientific or nonscientific discipline. Thanks to its rigid norms of truth, a discipline can pretend to entirely contain all knowledge within its own field. If the discipline in question is considered as fundamental, as a touchstone for all other disciplines, its scope is thereby enlarged so that it appears to encompass all human knowledge. In the classical viewpoint, the disciplines as a whole were conceptualized as a pyramid, the base of which was physics. Complexity literally pulverized this pyramid, provoking a veritable disciplinary big bang.

The fragmentation of the disciplinary universe is in full swing today. The domain of each discipline is inevitably becoming more and more specific; that which enables communication between disciplines is becoming more and more difficult, even impossible. A multischizoid, complex reality appears to have replaced the simple, one-dimensional reality of classical thought. In turn, the subject is put in shambles by being replaced with an ever-increasing number of separate parts, which are studied by the different disciplines. This is the price that the subject must pay for a certain kind of self-established knowledge.

There are multiple causes for this disciplinary big bang and these could well be the object of several scholarly works. But the fundamental cause is perhaps easy to discern: the disciplinary big bang is the response to the demands of a technoscience without brakes, without values, without any end other than utilitarianism.

But this disciplinary big bang also has enormous positive consequences because it has led to an unprecedented understanding of the knowledge of the exterior universe, as well as contributing new impetus to the establishment of a new world viewpoint. Since a stick always has two ends, when the weight shifts too far in either direction, the balance must be restored.

In a paradoxical way, complexity is embedded in the very heart of simplicity: fundamental physics. Indeed, popular works state that contemporary physics is a physics where a wonderful simplicity rules, through fundamental "building-blocks"—quarks, leptons, and messengers of the physical interactions. Each discovery of a new building block, predicted by this theory, is saluted by a Nobel prize and is presented as a triumph of the simplicity that rules in the quantum world. But for physicists working inside physics, the situation appears as infinitely more complex.

The founders of quantum physics were looking for particles that, insofar as they are fundamental building blocks, could describe all physical complexity. But by 1960 this dream had already been shattered due to certain particles, which had been discovered thanks to particle accelerators. A new simplification was proposed by the introduction of the bootstrap principle in strong interactions: there is a kind of nuclear democracy, in which all particles are as fundamental as all other particles and one particle is what it is because all the other particles exist at the same time. In turn, the vision of the self-consistency of particles and their laws of interaction, though fascinating on the philosophical plane, is crumbling because of the unparalleled complexity of equations necessary to explain this self-consistency and the practical impossibility of finding solutions. The introduction of the subconstituents of hadrons (particles having strong interactions)—quarks—went on to replace the bootstrap proposition and also to introduce a new simplification in the quantum world. This simplification has led to still greater simplification, which dominates particle physics today: research on the grand-unification theories and the superunification of physical interactions. But there, too, complexity was not long in asserting its omnipotence.

For example, according to the superstring theory in particle physics, physical interactions appear to be very simple, unified, and subordinate to general principles if they are traced within a multidimensional space–time continuum and involve an incredible energy, corresponding to Planck's mass. But complexity appears at the moment they pass to our world, which is characterized by four dimensions and by low energies. Unified theories are at their strongest at the level of general principles, but they are very poor at describing the complexity on our own level of reality. There are even some rigorous mathematical results that indicate that the passage from one single unified interaction to the four known physical interactions is extremely difficult, even impossible. A raft of mathematical and experimental questions of extraordinary complexity remain unanswered. In contemporary physics mathematical and experimental complexity are inseparable.

It is interesting to note in passing that the superstring theory has emerged thanks to string theory, which in turn emerged from the bootstrap theory. According to the string theory, hadrons are represented by vibrating strings, which carry quarks and antiquarks at each end. For example, a meson is represented by a string that is like a stick with two ends: a quark and an antiquark. It is impossible to separate the two ends of a string: when one cuts a string, one does not obtain a quark and an antiquark, but two new strings, which always have two ends. If someone is obsessed by the separation of two ends of a string, he runs into a theoretical impossibility: quarks and antiquarks remain imprisoned forever in the interior of hadrons. It would require infinite energy to lengthen and completely separate a quark and an antiquark. This paradoxical but nevertheless simple property in fact hides an infinite complexity in the interaction between quantum particles. Physicists still have not discovered a rigorous mathematical demonstration of the confinement of quarks.

Complexity is manifested everywhere else in the exact (hard) or human (soft) sciences. In biology and in neuroscience, for example, which are presently rapidly developing, each day brings more complexity and new surprises.

The development of complexity is particularly striking in the arts. By an interesting coincidence, abstract art appeared at the same time as quantum physics. But since then, an increasingly chaotic development seems to dominate more and more formalized experiments in this field. With some notable exceptions, such as Brancusi's sculpture, meaning has been wiped out by form. The human face, so beautiful during the art of the Renaissance, has been distorted more and more until it seems to have disappeared totally into absurdity and meaninglessness. A new art, computer art, has arisen and gradually has replaced aesthetic works and acts. In art, as elsewhere, a stick always has two ends.

Social complexity emphasizes the complexity that invades all areas of knowledge to the point of paroxysm. The simplistic idea of a just society, founded on a scientific ideology and the creation of the new man, is unraveling under the pull of a multidimensional complexity. That which remains is founded on the logic of utilitarianism and suggests to us nothing other than the "end of history." Everything happens as if there were no more future. And if there is no more future, sound logic tells us that there is no more present. The conflict between individual life and social life grows with an accelerating rhythm. And how can one dream of a social harmony based on the annihilation of the interior being?

Edgar Morin was right when he emphasized over and over that the knowledge of complexity demands a politics of civilization.

The knowledge of complexity, in order to be recognized as knowledge, bypasses one preliminary question: Is the complexity of which we speak a complexity without order, in which case its knowledge would have no meaning, or does it contain a new order and a new kind of simplicity which could appropriately become the object of a new knowledge? An authentic attempt to answer this question leads to a possibility between damnation and salvation.

Is complexity itself created by our minds or is it found in the very nature of things and beings? The study of natural systems gives us a partial answer to this question: both are true. Complexity in science is first of all the complexity of equations and of models. It is therefore the product of our mind, which is inherently complex. But this complexity is the mirror image of the complexity of experimental data, which proliferate endlessly; it is therefore also in the nature of things.

Still more, quantum physics and quantum cosmologies show us that the complexity of the universe is not the complexity of a garbage can, without any order. A stunning coherence exists in the relationship between the infinitely small and the infinitely large. A single term is lacking in this equation: our own finite realm, with its infinite dimensions. The subject remains silenced and estranged in the comprehension of complexity; and with reason, because he has been proclaimed dead. Between the two ends of the stick—simplicity and complexity—the included middle is missing: the subject himself.

7

A New Vision of the World— Transdisciplinarity

The process of the decline of civilizations is one of enormous complexity and its roots lie deep in obscurity. Of course, one can find multiple after-the-fact explanations and rationalizations without ever successfully dissipating the feeling that there is an irrational element at work at the heart of the process. Neither the masses nor great decision makers, as actors in a well-defined civilization, seem able to stop the decline of their civilization, even if they become more or less aware of the processes at work.

One thing is certain: this fall is always accompanied by a great imbalance between the mentalities of the actors and the inner developmental needs of a particular type of society. Although a civilization never stops proliferating new knowledge, it is as if these can never be fully integrated within those who belong to this civilization. And it is the human being who must be placed at the center of any civilization worthy of the name.

The unprecedented increase of knowledge in our era raises the challenging question of how to adapt our mentality to being. The challenge is enormous, because the influence of the Western-type civilization around the globe is so pervasive that its collapse would be even more devastating than the destruction we suffered in the two world wars.

Within the framework of classical thought, the only existing solutions for escape from this declining situation are social revolution or the return to a supposedly "Golden Age."

We've already tried social revolution and the results have been catastrophic. The New Man turned out to be only a sad, empty man. No matter what cosmetic alterations the concept of "social revolution" undergoes, they will not be able to erase from our collective memory what was experienced during the century just ended.

The return to the Golden Age has not yet been tried, for the simple reason that the existence of a Golden Age in the first place has not been established. Even if one supposes that a Golden Age existed in time immemorial, such a return would have to be accompanied by an inner revolution of dogmatism, the mirror image of the social revolution. The different religious fundamentalisms that plague the world currently are an evil portent of the violence and bloodshed that would burst forth from this caricature of authentic "inner revolution."

As always, there is a third solution, that which constitutes the object of the present manifesto.

Harmony between inner being and outer knowledge presupposes that these known facts would be intelligible, comprehensible. But can such comprehension exist in the era of the disciplinary big bang and relentless specialization?

In our time, a Pico della Mirandola is inconceivable. Today, even two specialists in the same discipline must make a serious effort to understand their respective results. There is nothing especially troubling about this insofar as it is the collective intelligence of the community attached to a certain discipline that leads to its progress, not simply a single brain that must necessarily know all the results of all his colleagues' brains—clearly an impossibility. Today there are hundreds of disciplines. How can a theoretical particle physicist truly hold a dialogue with a neurophysiologist, a mathematician with a poet, a biologist with an economist, a politician with a computer programmer, beyond mouthing more or less banal generalities? Yet, a true decision maker must be able to have a dialogue with all of them at once. Disciplinary language is an apparently insurmountable barrier for a neophyte, and each of us is a neophyte in some area. Is a modern Tower of Babel inevitable?

Perhaps a Pico della Mirandola in our time might be conceivable if he were to take the form of a supercomputer into which one could load all the known data that has been generated by all existing disciplines. This supercomputer would be capable of knowing everything while understanding nothing. Its user would be no better off than the supercomputer itself. The user would have immediate access to any results of any importance from any discipline, but would be incapable of understanding their meanings, still less of making connections between the results of different disciplines.

This process of "Babelization" cannot continue without putting our own existence into jeopardy, because a decision maker becomes increasingly more incompetent regardless of his or her intention. Without exception, each of the major challenges of our era—for example, the challenge of formulating an ethics adapted to the contemporary world—requires more and more competencies. However, it is obvious that even a group comprised of the best specialists from all the various disciplines would only be able to develop a generalized incompetence, for the simple reason that the sum total of competencies is not competence: on the technical level, the intersection between different domains of knowledge is an empty ensemble. Now, what is a decision maker, individual or collective, if not someone capable of taking into account all the givens of the problem that is being examined?

The indispensable need for bridges between the different disciplines is attested to by the emergence of multidisciplinarity and interdisciplinarity around the middle of the twentieth century.

Multidisciplinarity concerns studying a research topic not in just one discipline but in several at the same time. For example, a painting by Giotto can be studied not only within the context of art history, but also within the contexts of the history of religions, European history, or geometry. Marxist philosophy can be studied by blending philosophy with physics, economics, psychoanalysis, or literature. Any topic in question will ultimately be enriched by incorporating the perspectives of several disciplines. Moreover, our understanding of the topic in terms of its own discipline is deepened by a fertile multidisciplinary approach. Multidisciplinarity brings a plus to the discipline in question (the history of art or philosophy, in our examples), but we must remember that this "plus" is always in the exclusive service of the home discipline. In other words, the multidisciplinary approach overflows disciplinary boundaries while its goal remains limited to the framework of disciplinary research.

Interdisciplinarity has a different goal than multidisciplinarity. It concerns the transfer of methods from one discipline to another. One can distinguish three degrees of interdisciplinarity: (a) degree of application (for example, when the methods of nuclear physics are transferred to medicine, which leads to the appearance of new treatments for cancer); (b) epistemological degree (such as, transferring methods of formal logic to the area of general law, which generates some interesting analyses of the epistemology of law); (c) degree of the generation of new disciplines (when methods from mathematics are transferred to physics, generating mathematical physics, or when mathematical methods are transferred to meterological phenomena or stock market processes, generating chaos theory; transferring methods from particle physics to astrophysics produces quantum cosmology; and the transfer of computer methods to art, leads to computer art). Like multidisciplinarity, interdisciplinarity overflows the disciplines, but its goal still remains within the framework of disciplinary research. It is through the third degree that interdisciplinarity contributes to what we have called the disciplinary big bang.

As the prefix *trans* indicates, transdisciplinarity concerns that which is at once between the disciplines, across the different disciplines, and beyond all discipline. Its goal is the understanding of the present world, of which one of the imperatives is the unity of knowledge.

Is there something between and across the disciplines and beyond all disciplines? From the point of view of classical thought there is absolutely nothing. The space in question is empty, completely void, like the vacuum of classical physics. Even when the pyramidal vision of knowledge is renounced, classical thought considers each fragment of the pyramid that is generated by the disciplinary big bang as an entire pyramid; each discipline claims that it is sufficient unto itself. From the point of view of classical thought, transdisciplinarity appears absurd because it has no object. In contrast, within the framework of transdisciplinarity, classical thought does not appear absurd; it simply appears to have a restricted sphere of applicability.

In the presence of several levels of Reality, the space between disciplines and beyond disciplines is full, just as the quantum vacuum is full of all potentialities: from the quantum particle to the galaxies, from the quark to the heavy elements that condition the appearance of life in the universe.

The discontinuous structure of the levels of Reality determines the discontinuous structure of transdisciplinary space, which in turn explains why transdisciplinary research is radically distinct from disciplinary research, even while being entirely complementary. Disciplinary research concerns, at most, one and the same level of Reality; moreover, in most cases, it only concerns fragments of one level of Reality. In contrast, transdisciplinarity concerns the dynamics engendered by the action of several levels of Reality at once. The discovery of these dynamics necessarily passes through disciplinary knowledge. While not a new discipline or a new superdiscipline, transdisciplinarity is nourished by disciplinary research; in turn, disciplinary research is clarified by transdisciplinary knowledge in a new and fertile way. In this sense, disciplinary and transdisciplinary research are not antagonistic but complementary.

The three pillars of transdisciplinarity—levels of Reality, the logic of the included middle, and complexity—determine the methodology of transdisciplinary research.

There is an interesting parallel between the three pillars of transdisciplinarity and the three postulates of modern science.

In spite of an almost infinite diversity of methods, theories, and models that run throughout the history of different scientific disciplines, the three methodological postulates of modern science have remained unchanged from Galileo until our day. Only one science has entirely and integrally satisfied the three postulates: physics. The other scientific disciplines only partially satisfy the three methodological postulates of modern science. However, the absence of rigorous mathematical formulation in psychology, history of religions, and a multitude of other disciplines does not lead to the elimination of these disciplines from the field of science. At least for the moment, not even an exact science such as molecular biology can claim a mathematical formulation as rigorous as that of physics. In other words, there are degrees of disciplinarity that can respectively take into account more or less completely the three methodological postulates of modern science.

Likewise, the process of more or less completely taking into account the three methodological pillars of transdisciplinary research generates different degrees of transdisciplinarity. Transdisciplinary research, which corresponds to a certain stage of transdisciplinarity, will be closer to multidisciplinarity (as in the case of ethics); research that corresponds to another degree will be closer to interdisciplinarity (as in the case of epistemology); and that corresponding to yet another degree will be closer to disciplinarity.

Disciplinarity, multidisciplinarity, interdisciplinarity, and transdisciplinarity are like four arrows shot from but a single bow: knowledge.

As in the case of disciplinarity, transdisciplinary research is not antagonistic but complementary to multidisciplinary and interdisciplinary research. Transdisciplinarity is nevertheless radically distinct from multidisciplinarity and interdisciplinarity because of its goal, the understanding of the present world, which cannot be accomplished in the framework of disciplinary research. The goal of multidisciplinarity and interdisciplinarity always remains within the framework of disciplinary research. If transdisciplinarity is often confused with interdisciplinarity and multidisciplinarity, (and by the same token, we note that interdisciplinarity is often confused with multidisciplinarity), this is explained in large part by the fact that all three overflow disciplinary boundaries. This confusion is very harmful to the extent that it functions to hide the different goals of these three new approaches.

Although we recognize the radically distinct character of transdisciplinarity in relation to disciplinarity, multidisciplinarity, and interdisciplinarity, it would be extremely dangerous to absolutize this distinction, in which case transdisciplinarity would be emptied of all its contents and its usefulness in action reduced to nothing.

The complementary character of disciplinary, multidisciplinary, interdisciplinary, and transdisciplinary approaches is demonstrated in a stunning way, for example, by our attitude toward death: how we attend to the dying. This relatively new approach to the dying is extremely important because in recognizing the role of our death in our life, we discover hitherto unsuspected dimensions of life itself. The way we assist the dying is greatly enriched by transdisciplinary research because deeper understanding of the present world must pass through deeper understanding of the meaning of our life and of our death in this world which is ours.

8

Transdisciplinarity and the Open Unity of the World

The transdisciplinary viewpoint allows us to consider a multidimensional Reality, structured by multiple levels replacing the single-level, one-dimensional reality of classical thought. This proposal is not enough, by itself, to justify a new world viewpoint. We must first of all answer many questions in the most rigorous way possible. What is the nature of the theory that can describe the passage from one level of Reality to another? Is there truly a coherence, a unity among all levels of Reality? What is the role of the subject–observer of Reality in the dynamics of this possible unity? Is there a level of Reality that is privileged relative to all other levels? What is the role of reason in the dynamics of this possible unity of knowledge? What is the predictive power of the new model of Reality in the realms of reflection and action? Finally, is understanding the present world possible?

According to our model, Reality is structured via a certain number of levels. The considerations that follow do not depend on whether or not this number is finite or infinite. For the sake of clarity, let us suppose that this number is infinite.

Two adjacent levels are connected by the logic of the included middle in the sense that the T-state present at a certain level is connected to a pair of contradictories (A and non-A) at an immediately adjacent level. The T-state allows the unification of contradictories A and non-A, but this unification takes place at a level different from the one on which A and non-A are situated. The axiom of noncontradiction is thereby respected. Does this mean that we can obtain a complete theory, which can account for all known and forthcoming results? The answer to this question is not just theoretically interesting. After all, all ideologies and fanaticisms that claim changing the world is their aim are founded on the belief that their way is complete. Such ideologies and fanaticisms claim to possess the whole truth, the only truth.

There is certainly a coherence among different levels of Reality, at least in the natural world. In fact, an immense self-consistency—a cosmic bootstrap—seems to govern the evolution of the universe, from the infinitely small to the infinitely large, from the infinitely brief to the infinitely long. For example, one very slight variation of the coupling constant of the strong interactions between quantum particles could lead, on the macrophysical level—that of our universe—either to the conversion of all hydrogen into helium, or to the nonexistence of the complex atoms such as carbon. Or else a very small variation of the gravitational constant could lead either to some ephemeral planets or to the impossibility of their formation. Still further, according to current cosmological theories, the universe may be capable of self-creation without any external intervention. A flow of information is transmitted in a coherent manner from one level of Reality to another in our physical universe.

The logic of the included middle is capable of describing the coherence among these levels of Reality by an iterative process defined by the following stages: (1) A pair of contradictories (A, non-A) situated at a certain level of Reality is unified by a T-state situated at a contiguous level of Reality; (2) In turn, this T-state is linked to a couple of contradictories (A^1, non-A^1), situated at its own level; (3) The pair of contradictories (A^1, non-A^1) is, in its turn, unified by a T'-state situated at a third level of Reality, immediately contiguous to that where the ternary (A^1, non-A^1, T) is found. The iterative process continues indefinitely until all the levels of Reality, known or conceivable, are exhausted.

In other words, the action of the logic of the included middle on the different levels of Reality induces an open structure of the unity of levels of Reality. This structure has considerable consequences for the theory of knowledge because it implies the impossibility of a self-enclosed complete theory.

In effect, in accordance with the axiom of noncontradiction, the T-state realizes the unification of a pair of contradictories (A, non-A) but is associated, at the same time, with another pair of contradictories (A^1, non-A^1). This signifies that starting from a certain number of mutually exclusive pairs, one can construct a new theory which eliminates contradictions at a certain level of Reality, but this theory is only temporary because it inevitably leads, under the joint pressure of theory and experience, to the discovery of new pairs of contradictories, situated at new levels of Reality. In turn this theory will therefore be replaced by still more unified theories as new levels of Reality are discovered. This process will continue indefinitely without ever resulting in a completely unified theory. The axiom of noncontradiction is increasingly strengthened during this process. In this sense, without ever reaching absolute noncontradiction, we can speak of an evolution of knowledge, which encompasses all levels of Reality: knowledge that is forever open. In the sphere of levels of Reality per se, that which is above is like that which is below, but that which is below is not like that which is above. Finer matter penetrates coarser matter, just as quantum matter penetrates macrophysical matter, but the reverse is not true. Degrees of materiality induce an orienting arrow for tracing the transmission of information from one level to the other. In this sense, that which is below is not the same as that which is above; the words *high* and *low* here have no other meaning (spatial or moral) than that which is topologically associated with the flow of transmission of information. In its turn, this orienting arrow is associated with the discovery of more and more unifying and encompassing general laws.

The open structure of the unity of levels of Reality is in accord with one of the most important scientific results of the twentieth century concerning arithmetic, the theorem of Kurt Gödel, which states that a sufficiently rich system of axioms inevitably leads to results that are either indecisive or contradictory. The implications of Gödel's theorem have considerable importance for all modern theories of knowledge, primarily because it concerns not just the field of arithmetic, but all of mathematics, which is based on arithmetic. Obviously the mathematics that underlies theoretical physics includes arithmetic. This means that all research for a complete physical theory is illusory. If this affirmation is true in the most rigorous fields of the study of natural systems, how can one imagine a complete theory in an infinitely more complex sphere—that of the human sciences?

In fact, the search for an axiomatic system leading to a complete theory (without indecisive or contradictory results) represents at the same time both the apex and the beginning point of the decline of classical thought. The axiomatic dream is unraveled by the verdict of the holy of holies of classical thought: mathematical rigor.

The theorem that Gödel demonstrated in 1931 caused only a faint echo beyond a very limited circle of specialists. The difficulty and extreme subtlety of its proof explains why this theorem has taken so long to be understood within the mathematical community. Today it has scarcely begun to penetrate the world of physicists (though Wolfgang Pauli, one of the founders of quantum mechanics, was one of the first physicists to understand the extreme importance Gödel's theorem has for the construction of physical theories). Can one therefore blame Stalin for not having taken account of Gödel's theorem and, thus, for not having foreseen the fall of his empire?

The Gödelian structure of the unity of levels of Reality, associated with the logic of the included middle, implies that it is impossible to construct a complete theory for describing the passage from one level to the other, and, a fortiori, for describing the unity of levels of Reality.

If such unity does exist, this linking of all the levels of Reality must necessarily be an open unity.

To be sure, there is a coherence of the unity of levels of Reality, but this coherence is oriented in a certain direction: there is an arrow associated with all transmission of information from one level to the other. As a consequence of this, if coherence is limited only to certain levels of Reality, it stops both at the "highest" level and at the "lowest" level. If we wish to suggest the idea of a coherence that continues beyond these two limiting levels, so that there is an open unity, we must conceive the unity of levels of Reality as a unity that extends by a zone of nonresistance to our experiences, representations, descriptions, images, and mathematical formulations. Within our model of Reality, this zone of nonresistance corresponds to the "veil" which Bernard d'Espagnat referred to as "the veil of the real." The "highest" level and the "lowest" level of the totality of levels of Reality are united across a zone of absolute transparence. But these two levels are different, and therefore from the point of view of our experiences, representations, descriptions, images, and mathematical formulations, absolute transparence functions like a veil. In fact, the open unity of the world implies that that which is "low" is the same as that which is "high." The correspondences between "high" and "low" are established by the zone of nonresistance.

Quite simply, the nonresistance of this zone of absolute transparence is due to the limitations of our bodies and of our sense organs—limitations that apply regardless of what measuring tools are used to extend these sense organs. To claim that there is an infinite human knowledge (excluding any zone of nonresistance), while simultaneously affirming the limitations of our body and our sense organs, seems like linguistic sleight of hand. The zone of nonresistance corresponds to the sacred—to that which does not submit to any rationalization. Proclaiming that there is a single level of Reality eliminates the sacred, and that level unavoidably self-destructs.

The unity of levels of Reality and its complementary zone of nonresistance constitutes what we call the transdisciplinary Object.

A new Principle of Relativity emerges from the coexistence between complex plurality and open unity: no level of Reality constitutes a privileged place from which one is able to understand all the other levels of Reality. A level of Reality is what it is because all the other levels exist at the same time. This Principle of Relativity is what originates a new perspective on religion, politics, art, education, and social life. And when our perspective on the world changes, the world changes. In the transdisciplinary vision, Reality is not only multidimensional, it is also multireferential.

The different levels of Reality are accessible to human knowledge thanks to the existence of different levels of perception, which are found in a one-to-one correspondence with levels of Reality. These levels of perception permit an increasingly general, unifying, encompassing vision of Reality, without ever entirely exhausting it.

As in the case of levels of Reality, the coherence of levels of perception presuppose a zone of nonresistance to perception.

The unity of levels of perception and this complementary zone of nonresistance constitutes what we call the transdisciplinary Subject.

The two zones of nonresistance of transdisciplinary Object and Subject must be identical for the transdisciplinary Subject to communicate with the transdisciplinary Object. A flow of consciousness that coherently cuts across different levels of perception must correspond to the flow of information coherently cutting across different levels of Reality. The two flows are interrelated because they share the same zone of nonresistance. Knowledge is neither exterior nor interior: it is simultaneously exterior and interior. The studies of the universe and of the human being sustain one another. The zone of nonresistance plays the role of the secretly included middle which allows the unification of the transdisciplinary Subject and the transdisciplinary Object while preserving their difference.

The role of the explicitly or secretly included middle in the new transdisciplinary model of Reality is not so surprising after all. The words *three* (from the Latin *tres*) and *trans* have the same etymological root: three signifies "the transgression of the two, that which goes beyond the two." Transdisciplinarity transgresses the duality of opposing binary pairs: subject/object, subjectivity/objectivity, matter/consciousness, nature/divine, simplicity/complexity, reductionism/holism, diversity/unity. This duality is transgressed by the open unity that encompasses both the universe and the human being.

The transdisciplinary model of Reality has, in particular, some important consequences in the study of complexity. Without its contradictory pole of simplicity, complexity appears as an increasingly greater distance between the human being and Reality and introduces a self-destructive alienation of the human being thus faced with the absurdity of his own destiny. But the infinite simplicity of the transdisciplinary Subject corresponds to the infinite complexity of the transdisciplinary Object, just as the terrifying complexity of one level of Reality can signify the harmonious simplicity of another level of Reality.

The open unity between the transdisciplinary Object and the transdisciplinary Subject is conveyed by the coherent orientation of the flow of information, which cuts through the levels of Reality, and of the flow of consciousness, which cuts through the levels of perception. This coherent orientation gives a new, deeper meaning to the simple fact of human verticality in the world. Instead of individual human verticality ruled by the law of universal gravity, the transdisciplinary viewpoint posits a conscious and cosmic verticality, traversing the different levels of Reality. In the transdisciplinary vision, it is this cosmic verticality that constitutes the foundation of any viable social project.

9

The Death and the Resurrection of Nature

Modernity is particularly deadly. It has invented all kinds of "deaths" and "ends": the death of God, the death of man, the end of ideologies, the end of history. But, there is a death that is spoken of much less, perhaps due to shame or ignorance: the death of Nature. In my view, this death of Nature is the source of all the other deadly concepts just invoked. In any case, the very word *Nature* has ended up disappearing from scientific vocabulary. Of course, the layperson, like the scientist (in popular writings), still uses this word, but in a confused, sentimental way, reminiscent of magic. How have we arrived at this state of affairs?

Since the beginning of time we have not stopped modifying our vision of Nature. Historians of science agree that, despite all appearances to the contrary, there has not been just one vision of Nature throughout time. What can there be in common between the Nature of so-called primitive peoples, the Nature of the Greeks, and the Natures of Galileo, the Marquis de Sade, Laplace, or Novalis? Nothing, except the human being himself: the view of Nature of a given period depends on the kind of imagination that predominates during that period; in turn, that view depends on many factors: the degree of scientific and technological development, social organization, art, religion, and so forth. Once formed, an image of Nature exercises an influence on all areas of knowledge. The passage from one viewpoint to another is not progressive, continuous—it happens by sharp, radical, discontinuous breaks. Several contradictory viewpoints can coexist. The extraordinary diversity of viewpoints of Nature explains why one cannot speak of Nature, but only of a certain nature in accord with the imagination of a given period.

It is important to emphasize that the idea of a privileged, if not exclusive, relationship between Nature and science is a recent prejudice, founded upon the scientistic ideology of the nineteenth century. The historical reality is much more complex. The image of Nature has always had a multiform action: it has influenced not only science but also art, religion, and social life. This allows us to explain some strange synchronicities. Here I limit myself to a single example: the simultaneous appearance in the past few years of the theory of the end of history and of superunified theories in particle physics. These physical theories have as their aim the elaboration of a complete approach, founded on a unique interaction, which can predict everything (hence the name, "Theory of Everything"). It is quite obvious that if such a theory were actually formulated in the future, it would signify the end of fundamental physics, because there would be nothing important left to look for. It is interesting to observe that both the idea of the end of history and of the end of physics have simultaneously emerged from the "end of the century" imagination. Is this mere coincidence?

Notwithstanding the abundant and fascinating diversity of images of Nature, we can nevertheless distinguish three main groupings: magic Nature, Nature as machine, and the death of Nature. The magical viewpoint depicts

Nature as a living organism, endowed with intelligence and consciousness. The fundamental postulate of such thought is one of universal interdependence: Nature cannot be conceived outside of its relationship to us. Everything is sign, trace, signature, symbol. Science, in the modern sense of this word, is superfluous.

At the other extreme, the mechanistic thought of the eighteenth and above all the nineteenth centuries (which, by the way, still predominates today) conceives Nature not as an organism, but as a machine: One only has to disassemble this machine piece by piece in order to possess it entirely. The fundamental postulate of mechanistic thought is that Nature can be known and conquered by scientific methodology, defined in a way that is completely independent of human beings and separate from us. The triumphant vision of "the conquest of Nature" is rooted in the formidable technical and technological usefulness of this postulate.

Certain scientists, artists, and philosophers have been profoundly affected by the deadly danger of mechanistic thought. Hence, there arose the antagonistic current of German *Naturphilosophie* (inspired by the work of Jacob Boehme), which revolved around the journal Athenaeum, and includes the writings of Schelling, Schlegel, Novalis, Ritter, and Goethe. Viewed from our own time frame, *Naturphilosophie* may seem a grotesque distortion, a crude manipulation of science—a sort of dead-end street, as if this were a ridiculous attempt to return to magical thought and to living Nature. But we cannot deny that this philosophy of Nature generated at least two major scientific discoveries: cellular theory and electromagnetism. I believe that the true fault of *Naturphilosophie* was that it appeared two centuries too soon, thereby missing the threefold quantum, technological, and computer revolutions.

The logical end result of the mechanist viewpoint was the death of Nature—the very disappearance of the concept of Nature from the scientific field. From the very beginning of the mechanistic vision, Nature as machine, with or without the image of God as watchmaker, fell to pieces. From that moment on, there was no more need for a coherent whole, for a living organism, or even for a machine that still kept the musty odor of finality. Nature was dead, but complexity remained: an astonishing complexity, which penetrated each and every field of knowledge. Yet this complexity was perceived as an accident; we ourselves are considered an accident of complexity—an altogether joyless viewpoint, which brings us to the world of today.

The death of Nature is incompatible with a coherent interpretation of the results of contemporary science, in spite of the persistence of the neoreductionistic attitude that accords an exclusive importance to the fundamental building blocks of matter and to the four known physical interactions. According to this neoreductionist attitude, all recourse to Nature is superfluous and devoid of sense. But regardless of the origin of the resistance implied in such reactionary attitudes, the time for the resurrection of Nature has come. In truth, Nature is dead only for a certain viewpoint of the world: the viewpoint of classical thought.

The rigid objectivity of such classical thought is not viable in the quantum world. The idea of a complete separation between observer and Reality assumed to be completely independent from that observer brings us to the verge of insurmountable paradoxes. In fact, a far more subtle notion of objectivity characterizes the quantum world: objectivity depending on the level of reality in question.

The empty vacuum of classical physics is replaced by the full vacuum of quantum physics. The smallest region of space is enlivened by astonishing activity, signifying perpetual movement. The quantum fluctuations of the void determine the sudden appearance of virtual particle/antiparticle pairs, which are annihilated in the course of extremely short intervals of time. Everything takes place as if the quantons of matter were created out of nothing. A metaphysician might say that the quantum void is a manifestation of one of the faces of God: God the Nothing. In any case, in the quantum vacuum, all is vibration, everything is a fluctuation between being and nonbeing. The quantum vacuum is full, filled with all potentialities, from the particle to the universe.

By supplying energy to the quantum vacuum we can help it to materialize its potentiality. This is precisely what we do when we build particle accelerators. When certain threshold energies are attained, real particles suddenly materialize at that point—being literally drawn out of nothing. These particles have an artificial character, in the true sense of that word. From our perspective, the macrophysical world appears to be constructed in an extremely economic way: protons, neutrons, and electrons are enough to build almost all of our visible universe. But scientists have succeeded in creating hundreds of other particles from nothingness: hadrons, leptons, electro-weak bosons, etc.

Space–time itself no longer remains a fixed concept. Our space–time, involving four dimensions, is not the only conceivable space–time. According to certain physical theories, it appears more like an approximation, like a part of a space–time that is richer as generator of possible phenomena. Supplementary dimensions are not the result of mere intellectual speculation. On the one hand, these dimensions are necessary to ensure the self-consistency of the theory and the elimination of certain undesirable aspects. On the other hand, they do not have a purely formal character—they have physical consequences for our own scale. For example, according to certain cosmological theories, if the universe had been associated at the "beginning" of the big bang with a multidimensional space–time, supplementary dimensions would have remained forever hidden, unobservable; instead, their vestiges would be precisely those known physical interactions. By generalizing the example provided by particle physics, we can conceive that certain levels of Reality correspond to a space–time different than that characterizing our own level. Moreover, complexity itself must depend on the nature of space–time as well.

According to actual scientific conceptions, matter is far from being identical with substance. In the quantum world, we observe a perpetual energy–substance–information transformation; the concept of energy appears like a unifying concept: information is coded energy, while substance is concretized energy. In contemporary physics, space-time itself does not appear to be like a receptacle into which material objects are plunged: it is a consequence of the presence of matter. Matter is associated with a substance–energy–information–space–time manifold. The degree of quantum materiality is, of course, different from the degree of materiality considered by classical physics.

The nature of complexity changes. It is no longer a complexity directly reducible to simplicity. Different degrees of materiality correspond to different degrees of complexity: what can be conceived as extreme complexity from the perspective of one level of Reality may be seen as simplicity relative to another level of Reality, but exploration of this second level reveals that this simplicity corresponds to extreme complexity relative to its own laws. This structure of degrees of complexity is intimately tied to the Gödelian structure of Nature and of knowledge, which is induced by the existence of different levels of Reality.

The same notion of laws of Nature completely changes when compared to the classical viewpoint. This situation can be summed up by three theses formulated by the physicist Walter Thirring:

1. The laws of any inferior level are not completely determined by the laws of a superior level. Thus, notions well-anchored in classical physics, such as "fundamental" and "accidental," must be re-examined. That which is considered to be fundamental on one level may appear accidental on a superior level, and that which is considered to be accidental or incomprehensible on a certain level can appear to be fundamental on a superior level.

2. The laws of an inferior level depend more on the circumstances of their emergence than do the laws of a superior level. The laws of a certain level depend essentially on the local configuration to which these laws refer. There is therefore a kind of local autonomy of respective levels of reality; however, certain internal ambiguities concerning laws of an inferior level of reality are resolved by taking into account the laws of a superior level. It is the internal consistency of laws that reduces the ambiguity of laws.

3. The hierarchy of laws evolved at the same time as the universe itself. In other words, the birth of laws occurs simultaneously with the evolution of the universe. These laws pre-exist at the "beginning" of the universe as potentialities. It is the evolution of the universe that actualizes these laws and their hierarchy. A transdisciplinary model of Nature must integrate all these new characteristics of the physical universe.

We can distinguish three major aspects of Nature according to the transdisciplinary model of Reality:

1. Objective Nature, which is connected with the natural properties of the transdisciplinary Object; objective

Nature is subject to subjective objectivity. This objectivity is subjective to the extent that the levels of Reality are linked to levels of perception. Nevertheless, the emphasis here is on objectivity, to the extent to which the methodology employed is that of science.

2. Subjective Nature, which is connected with the natural properties of the transdisciplinary Subject; subjective Nature is subject to objective subjectivity. This subjectivity is objective to the extent that the levels of perception are connected with levels of Reality. Nevertheless, the emphasis here is on subjectivity, to the extent to which the methodology employed is that of the ancient science of being, which is present in the traditions and religions of the world.

3. Trans-Nature, which is connected with a similarity in Nature—a veritable communion—that exists between the transdisciplinary Object and the transdisciplinary Subject. Trans-Nature concerns the domain of the sacred and corresponds to the "veil" which is the zone of nonresistance mentioned in the previous chapter. It cannot be approached without considering the other two aspects of Nature.

Transdisciplinary Nature has a ternary structure (objective Nature, subjective Nature, trans-Nature), which defines living Nature. This Nature is living because it is there that life is present in all its degrees and because its study demands the integration of lived experience. The three aspects of Nature must be considered simultaneously, in terms of their interrelation and their conjunction within all the phenomena of living Nature. The study of living Nature asks for a new methodology—transdisciplinary methodology—which is different from the methodology of modern science and from the methodology of the ancient science of being. It is the coevolution of the human being and of the universe which asks for a new methodology. The fullness of living Nature gives us a sense of something that could, sooner or later, become a transdisciplinary ecology.

An attempt to elaborate a new philosophy of Nature, a privileged mediator of a dialogue among all the areas of knowledge, is one of the highest priorities of transdisciplinarity.

The definition of Nature that I propose signifies neither a return to magical thought nor a return to mechanistic thought, because it rests on a twofold affirmation: (1) the human being can study Nature by means of science; (2) Nature cannot be conceived except in terms of its relation to the human being.

In fact the term *living Nature* is a pleonasm, because the word *Nature* is intimately linked to *birth*. The root of the Latin word *natura* is *nasci* and designates the action of giving birth as well as the feminine organs of generation. Living Nature is the womb of the self-engendering of the human being.

Galileo had a vision that Nature was like a text written in mathematical language which we can decipher and read. This vision, which has come down to us over the centuries, has shown itself to be one of enormous utility. However, today we know that the situation is far more complex. Nature seems more like a book in the process of being written: the book of Nature is therefore not so much to be read as experienced, as if we are participating in the writing of it.

Human beings have always dreamed of pondering their own face in the mirror of Nature. The mirror of magical Nature is, of course, a magic mirror: everything can be seen, perceived, experienced, at least potentially, in this mirror. Unity is actualized; diversity is potentialized.

In contrast, the mirror of mechanistic Nature is like a broken mirror, or a scalpel. It is enough to take one piece of the tissue (that is, Nature) in this mirror/scalpel in order to make a pronouncement about the entire Nature machine. This piece of Nature is conceived as if it were a miniature copy, conforming to the entire whole. The privileged instrument for interpreting the image produced by this mirror/scalpel is theory, more and more formalized on the mathematical plane. Etymologically, *theory* means the action of observing, the fruit of intellectual contemplation, the action of seeing a spectacle, or of participating in a feast. For mechanistic thought, the feast is transformed into a conquest and the spectacle is transformed into the reading of a book written in advance, the book of Nature. It is of little importance for whom or why this book has been written; from the moment it becomes entirely accessible, doors of unlimited power open before us.

The transdisciplinary Nature mirror is situated simultaneously between and beyond all areas of knowledge. The classical world is the world of figuration, while the transdisciplinary world is that of transfiguration. The old portrait of Nature is replaced by the icon of Nature. The word *mirror* comes from the Latin *mirare,* meaning "to look at with astonishment." The act of "looking at" presupposes two terms: the one who looks and the one who is being looked at. Where else could this astonishment come from if not from the included third?

In *The Conference of the Birds,* Attar, a Persian poet of the twelfth century, describes a long voyage of birds in search of their true king, the Simurgh. The birds cross seven valleys full of dangers and of wonders. The sixth valley is that of "astonishment." There it is at once day and night, one sees and one doesn't see, one exists and one does not exist, things are empty but at the same time they are full. At the end of their arduous journey, the birds find a mirror in which they are finally able to be seen and to be recognized.

10

Homo Sui Transcendentalis

With the evolution of tools and measuring devices, we observe a spectacular example of the relationship between the levels of human perception and those of physical Reality.

Since man first appeared on earth, he has invented tools for food gathering and protection from a hostile environment. These tools represented a genuine extension of the sense organs of the human body, but in the beginning this extension was limited to exploring the area immediately around the individual's own body.

In time, however, the human being discovered that he was able to transport his own body, carrying it far beyond the distance prescribed by his bodily limits. At first this transporting was on a horizontal plane, conforming to the law of gravity which holds the body to the earth, but humans dreamed of flying and thus liberating themselves from the restrictions of the earth's gravity. (Is it a coincidence that Icarus, the Greek hero who tried to fly, was the son of Dedalus, the inventor of the earth-formed labyrinth created to imprison the Minotaur?) It has been scarcely a century since this dream of flying has been realized: transport has become vertical as well, with airplane flights and interplanetary rockets allowing us to travel through different levels of Reality.

Today, the relentless desire of the human being to transgress or go beyond his own body leads, via genetic manipulation, to the potential transformation of his genetic memory, inherited from the age-old adventure of the planet Earth's giving birth to the human being's body.

But it is the human field of vision that undergoes a very radical change through technoscientific extension. The transgression of the limitations of sight has been accelerated by the appearance of astronomical lenses: the telescope. Galileo turned these lenses toward the sky and in a few months discovered a new world opening before his astonished eyes. The giant telescopes of today have only increased this accelerating exploration on the level of the infinitely large.

In the other direction, that of the infinitely small, an unexpected fact seemed to limit this transgression of the field of sight: microscopes encountered the quantum barrier. Strictly speaking, quantum particles are invisible because they are nonlocalizable. But the inventive capacity of man is inexhaustible. He has invented exploratory instruments for this apparently forbidden world. Particle accelerators are for the quantum world what microscopes and telescopes are for the classical world. Particles indicate their presence by the number of pulses recorded by electronic computers. Their properties are electronically reconstructed and quantum laws are verified with greater and greater precision. The discovery of the new quantum world is an event comparable to the discovery of a new celestial world at the time of Galileo. Another horizon opens on the infinitely small. The transgression of the field of vision leads to a transvision: a new level of Reality that can be explored by means of science. Exploration prior to the quantum world went from the visible toward the visible, whereas now it goes from the visible toward the invisible: toward that which is beyond the visible.

The comprehension of this new level of Reality rests on a double perception: certainly exterior perception, thanks to the quantum particles that are moving in the accelerators, as active as if they were truly "drills" inside the quantum world, but also interior perception, the manifestation of what one can call the quantum imagination.

Since we are not quantum entities, we cannot ourselves travel to explore this quantum world, but we can nevertheless perceive it if we make the effort to integrate within ourselves the paradoxical information that is provided to us by scientific theory and experiments. First of all, this effort must penetrate an interior silence, by quieting habitual thought based on the perception of macrophysical levels. Habitual thought is very talkative; it incessantly tells us what is true and what is false and perpetually fabricates images adapted to our macrophysical level. How can we perceive the unity of contradictories if habitual thought speaks to us of absolute truth and absolute falsehood? How can we imagine discontinuity if our habitual thought tells us that it would be like trying to climb a ladder with completely disconnected steps? How can we feel what "nonseparability" means if habitual images tell us that everything—in this world—is separate? To silence habitual thought also means to abolish the throng of macrophysical images that accompany it. In this moment of silence—considered misleading and destabilizing by habitual thought—we discover within ourselves a level of natural perception of the unity of contradictories. Just as the quantum world is enclosed within the macrophysical world, so this new degree of perceptions is enclosed within our habitual perception. This is why very young children consider what is told in fairy tales to be normal; their perception of the included third has not yet had time to be submerged by the endless accumulation of information brought by exploring the macrophysical world, that is to say, by our normal life. Recent scientific observations show that infants have a global perception of their environment: for them it is nonseparability that is natural and separability that must be painfully learned. Nevertheless, they experience another kind of thinking, which precedes conceptual thinking.

In one sense, at the door of the quantum world, we can become again like infants by sacrificing our habits of thought, our certainties, our imagery, because the quantum imagination is an imagination without imagery. In this way, a veritable transfiguration is processed: beyond macrophysical images, another sphere of Reality is offered to our understanding.

Comprehension of the quantum world therefore passes through a lived experience that integrates knowing based on scientific theory and experiment into our very being, while making us discover a new level of perception within ourselves. Thus, the word *theory* recovers its etymological meaning, that of "contemplation."

The discovery of harmony between a level of perception and a level of Reality is crucial for our behavior in normal life. Without this discovery, macrophysical thought takes over the new level of Reality by reducing it to its own norms, by mutilating it through a manipulation, the consequences of which can only be harmful.

In a sense we become like Prometheus who stole the fire of heaven (and whose name signifies one who predicts): We have discovered the fire hidden in the heart of the atom. But then, Pandora was sent to Earth by Zeus to seduce Prometheus's brother Epimetheus, whose name signifies one who reflects too late. We also are in the position of Epimetheus: By unleashing atomic fire, we have opened Pandora's box. We stand caught between Prometheus and Epimetheus, between one who predicts and one who reflects too late: we are obliged to find the right balance between the two—that of one who understands and acts.

The harmony of the transdisciplinary Subject and the transdisciplinary Object is linked to the harmony between the levels of perception and levels of Reality. In the transdisciplinary vision, the classic real/imaginary dichotomy also disappears. We can think of a level of Reality as a fold of all levels of perception; and we can think of a level of perception as being a fold of the totality of levels of Reality. The real is a fold of the imagination and the imagination is a fold of the real.The ancients were right: there is indeed an *imaginatio vera*, a foundational, true, creative, visionary imagination.

From fold to fold, we invent ourselves.

Different levels of comprehension result in the harmonious integration of the understanding of different levels of Reality and of the understanding of different levels of perception. Reality being multiple and complex, the levels of comprehension are multiple and complex. But, since Reality is also an open unity, the different levels of comprehension are linked to each other within a single, open Whole, which includes both the transdisciplinary Subject and the transdisciplinary Object. This Whole opens onto the area of nonresistance of the sacred, which is common to the Subject and to the Object. Paradoxically, this area, an area of nonresistance in which Subject and Object are considered as separate, appears like an area of absolute resistance in which Subject and Object are unified, because this area resists all comprehension, no matter what its level. It is the harmony between levels of Reality and levels of perception that transforms nonresistance into absolute resistance. The sacred acquires a status of Reality in the same way that the levels of Reality do, without however constituting a new level of Reality, because it eludes all knowing. Between knowing and comprehension there is being. Yet, the sacred does not oppose reason: to the extent to which it ensures harmony between Subject and Object, the sacred is an integral part of the new rationality.

Reality encompasses the Subject, the Object, and the sacred, which are three facets of one and the same Reality. Without any one of these three facets, Reality is no longer real, but a dangerous phantasmagoria.

Reality reduced only to the Subject has caused traditional societies to be swept away by modernity. Reality reduced only to the Object has led to totalitarian systems. Reality reduced to the sacred has led to fanaticism and religious fundamentalism. A viable society joins these three aspects of Reality in a balanced way.

The emergence of the idea of levels of understanding paves the way for what could be the evolution of modern man.

We are only beginning to explore different levels of Reality linked to different levels of perception. This exploration marks the beginning of a new stage of our history, based on the knowledge of the exterior universe in harmony with the self-knowledge of the human being.

Respect for the trans-Nature of human nature implies the recognition in every human being of his double interior and exterior transcendence. This transcendence is the basis of our freedom. The transdisciplinary vision is incompatible with any attempt to reduce the human being to a definition that has some formal structure no matter what it is. Every human being is free to open himself, in his own way and through his own liberating self-transformation, to the self-knowledge of his spiritual destiny. The right to this meaning must be inscribed among the rights of man.

We have the choice between evolving or disappearing. Our evolution is a self-transcendence. No one is obliged to evolve. The natural constraints of the environment that have obliged man to evolve biologically are no longer exercised. Biological evolution has reached full term. A new kind of evolution is emerging linked to culture, science, consciousness.

Individual evolution and social evolution condition each other. The human being nourishes the being of humanity, and the being of humanity nourishes the human being. If individual evolution is conceivable even in the absence of social evolution, on the other hand, social revolution is unthinkable without individual evolution. It is precisely the orientation of the flow of consciousness cutting across the different levels of perception that gives meaning—meaning and direction—to this co-evolution. There is hidden here an aspect of democracy that merits profound study in all its dimensions. The challenges of all kinds—be they irrational conflicts that permeate social life, the destructive conflicts that menace the life of nations, or the danger of the self-destruction of our own species—can be solved if this individual and social co-evolution is respected.

The self-birth of the universe and the self-birth of humanity are inseparable. Science and consciousness, the two pillars of the future universal democracy, sustain each other. Science without understanding is the ruin of the human being, but consciousness without science is also his ruin. The responsibility of self-transcendence—our responsibility—is the included third that unites science and consciousness.

Homo sui transcendentalis is in the process of being born. He is not some new man but man reborn. This new birth is a potentiality inscribed in our very being.

Transgression originally meant "to pass to the other side, to cross." In time, the word began to mean, for the translators of the Bible "the violation of divine law"; for jurists "the violation of a law." Does crossing from one level of Reality to another, or from one level of perception to another, mean an infraction with respect to divine or human laws? No. Transdisciplinarity is a generalized transgression which opens an unlimited space for freedom, understanding, tolerance, and love.

11

Techno-Nature and Cyberspace

The last frontier of our body—that of our own brain—is about to be transgressed.

Human mentation is projected beyond itself, generating results that are not the products of so-called natural processes. These results of the advance of technoscience, beginning with the conquest of space and the first step of man on the moon and ending with Virtual Reality, have produced a veritable Techno-Nature which now coexists with cosmic processes that have been developing since the dawn of time, even before the appearance of humans. The latest result to emerge from this Techno-Nature is cyberspace, which has a unique role because it reaches a new barrier of human intelligence—the barrier of light. Signals multiply in this new space at the speed limit allowed by Nature—the speed of light.

The word *cyberspace* is polysemantic and can therefore lend itself to multiple misinterpretations. Sometimes it refers only to Virtual Reality, with the "information highway" and the Internet appearing to be unrelated phenomena. This is why it is preferable to introduce a new name—Cyber–Space–Time (CST)—to designate the whole of computer space, which is about to envelope the entire earth.

It is therefore appropriate to question the nature of this space–time. Is it really new or does it coincide with the space–time examined by physics? How many dimensions are there in CST? What is the logic that rules CST? Is the nature of CST material or immaterial? What is the place of the human being in CST? Does CST play a role in the evolution of humanity and of individual humans or does it correspond to their involution? Is it simply a fashionable phenomenon, or does it signify the emergence of a new level of Reality?

First of all, CST is natural and artificial at the same time.

CST is natural because its source is natural: the quantum world. As a matter of fact, the symbols used in its basic coding—0 and 1—actually denote quantum processes. Broadly speaking, 0 and 1 signify "open door–closed door" in the quantum word. They are already a "translation" in mathematical language of microphysical processes. Thus, 0 and 1 are mere metanumbers. However, the fundamental language is that of the quantum world; it is therefore the language of Nature, and is, by definition, universal.

At the same time, CST is artificial. First of all, the language used—that of mathematics—is artificial, beginning with the basic coding (0,1) and ending with more and more elaborate mathematical equations that are the seeds of an infinity of images most of which have no correspondence to the natural world. Thus, as in the quantum world, abstraction is not a tool for describing reality, but an inseparable component of reality. CST is also artificial because it arises from a sophisticated technology, carried out by human beings.

This natural–artificial double aspect poses a very serious question about a new interface between man and computer. Ultimately, this new interface takes place between man and Nature, which again raises the question of a third force, reconciling man and Nature.

A long history of human intelligence has been journeyed from the calculi of the Sumerians up to the supercomputers of today. The calculi were terra-cotta objects whose size and shape were precisely connected with a numbering system. These objects were sealed up in spherical clay jars, allowing the positive identification of the goods of each proprietor. Modern supercomputers have replaced Sumerian accounting jars; the binary code (0,1) has replaced the calculis, and electromagnetic waves, the human hand.

In spite of these dizzying advances in power and calculating abilities, CST is of a material nature.

The information that circulates in CST is every bit as material as a chair, a car, or a quantum particle. Electromagnetic waves are just as material as the earth from which the calculi were made: it is simply that their degrees of materiality are different. The term *the civilization of the immaterial* is misleading because it presupposes the identification of matter with substance. In modern physics matter is associated with the complex relationship: substance–energy–information–space–time. The semantic shift from material to immaterial is not merely naive, for it can lead to dangerous fantasies.

CST engenders a new relation of transformation: that between mathematical equations and images.

A veritable transformation from the real to the imaginary becomes possible as well. The substitution of material money (paper or metal) by electronic money is only a gross illustration of what is a far more generalized transformation. An essential characteristic of CST is its maximal capacity for interactions between the real and imaginary, between the concrete and the abstract, between the human body and mathematical equations. In principle, therefore, CST is able to reveal a new level of perception.

Finally, CST is characterized by the fact that its signals circulate at the speed limit of the natural world, "c," or the speed of light.

By itself, speed c is not anything extraordinary. We can still observe stars that have disappeared long ago from the sky simply because their light is just now reaching us, conveyed by speed c. The particles in the atoms of our body turn with speed c. What is new is the fact that the human being has created a space–time in which all speeds are equal to c. CST has a cosmic dimension—that of the planet Earth. One can even ask if CST is the same everywhere in the cosmos because, according to current knowledge, matter is the same throughout the whole universe.

What is the logic governing CST?

Superficially, while observing that the coding (0,1) is binary, we are able to believe that it is simply a question of classical, binary logic. The computer would thus be considered as a machine, albeit perfected, and incapable of interaction with the human being.

Three remarks show us that this conclusion is false:

1. One cannot confuse coding and logic. It is as if just because we can write included middle in the language of letters (i-n-c-l-u-d-e-d . . .), we could conclude that the "included middle" must submit

to the rules of the excluded middle, which is an obvious absurdity.

2. The source of CST is the quantum world, which is ruled by a logic different from classical logic (for example, the logic of the included middle).

3. Bringing the human body into CST alerts us to a new level of perception (basically that of the encounter with the "light barrier"), which reveals a world that breaks radically with the macrophysical world in which we pass our life. This "new world" is not ruled by classical logic: the chain of causes and effects is suspended, linear causality is abolished, discontinuity is no longer merely thought, but lived.

How to move about in CST requires a new type of surfing, a navigation in the bowels of nature and in interaction with ourselves. It is the source of a new kind of imagination that affects perception, which in turn feeds the imagination. There is a kind of loop that is created between the quantum imagination and the navigation in CST. Quantum processes play a certain role in the functioning of memory and consciousness. It is as if a mirror is revealed between quantum processes of the human brain and quantum processes of CST. For the first time in history, there is a possibility of the integration of the finite where we are, in the unity between the infinitely small and the infinitely large. To the extent that this "finite" is like a crystal in which consciousness is infinitely reflected, we assist, perhaps, in giving birth to the first historical type of ternary interaction (infinitely small, infinitely large, infinitely conscious). Here, there is an ontological chance, which can obviously be easily ruined, squandered, missed, wasted if it is not recognized for what it is.

With the discovery of the quantum world and cybernavigation, *homo sui transcendentalis* begins his adventure.

Yet in recent times we are witnessing the appearance of strange phenomena and some rather odd characters.

Self-proclaimed messiahs, lacking any sense of real Annunciation, promote the joys of the global village. Utopians who are far from Utopia and humanists far from humanism claim a fellowship without frontiers on the Internet. Disguised as high-priest sponsors of the absolute, businessmen in search of the absolute market suggest that we surf through the fabulous space–time of the Virtual-Reality Virgin. They celebrate masses sung on the altar of the planetary hypermarket. On the other hand, an army of prophets of doom brandish for us the vision of the countless dangers in this new world. Some theologian–astrophysicists far from God offer us an exalted dogma of spirit as a program, with love as a subprogram, with God himself seen as at last rational and tangible—through the filling up of cosmic space with cybernetic tissue.

The warnings against cyberspace multiply endlessly with an acceleration comparable to that of the expansion of cyberspace itself. This process is entirely natural. To a great extent, the warnings are the only means of defense by an ancient system, which is trying, at any cost, for its own profit, to swallow up novelty like a phagocyte.

In fact, witness the inevitably paradoxical and troubling birth of a new level of Reality.

The components of Techno-Nature, Cyber–Space–Time included, possess a particular characteristic: self-movement. Self-movement in Techno-Nature implies submission to a principle of maximalization: everything that can be made will be made. This principle of maximalization can lead to the most monstrous abuses, but it also has an immense creative potential. It is our responsibility to respond to the evolutionary possibility that is offered to us playing again the role of the included third.

Causality in CST is different from the local one ruling the macrophysical level or the global one ruling the quantum level. Causality in CST is a causality in an open loop, ruling the man–computer interface. Thanks to his interaction with the computer, the human being has discovered in himself a new level of perception and the computer has deepened its potential by interacting with the human being. A chimerical being, such as the Minotaur, with the body of a man and the head of a bull, could be born from this double recursive interaction and could menace our existence. However, we can also envisage an unprecedented freedom from the many constraints of daily life, transferring these constraints to Cyber–Space–Time, which reveals itself as a true machine to liberate time. The time gained can be dedicated to our own interior development.

The idea of a correspondence between psychical processes and microphysical processes is found in the writings of Korzybski, Jung, Pauli, and Lupasco. This concept is currently passing from the realm of theoretical speculation to that of practical application. This correspondence is the source of what may be the worst or the best in the emergence of CST in the planetary life. We have an immense responsibility: it is not a question of finding a solution to the increasingly complex problems that appear endlessly in our actual system of reference, but of changing the system of reference itself, by introducing a new way of understanding the dialectic between simplicity and complexity.

Cyber–Space–Time is neither deterministic nor indeterministic. It is the space of human choice. To the extent that CST allows bringing into play the notion of levels of Reality and the logic of the included middle, it is potentially a transcultural, transnational, and transpolitical space.

The choice that confronts us at first appears binary: a choice between absolute market and Cyber-Utopia, between an "era of merchants" and an "era of travelers." To paraphrase Antonio Machado, I would say that choosing between them is not the way: it is by free movement that the way is created.

But the stick always has two ends.

One end of the "global village" stick corresponds to a demagogic formula that camouflages a new form of the world's domination by the rich. The rich will become richer and richer and the poor, poorer and poorer. This is what I call "the era of merchants."

The other end of the "global village" stick corresponds to the possible emergence of a village of villages (like what is called a system of systems). Megalopolises—huge centers of the concentration of information—obviously become useless in CST. Megalopolises can be transformed into immense archival centers or museums. Large cities as a source of ugliness and violence would hence disappear. This village of villages could also become a welcoming place for transreligion, transculture, transpolitics. One immediate priority will be to recognize CST at the level of international law as a transnational space, a space that does not belong to anyone. Hence the necessity not only for equality of access but also for completely free access (or circulation) in CST. In sum, this is what I call "the era of travelers."

Is the era of travelers in opposition to the era of merchants? No, not if each end of the stick finds its own place by not assuming it is the whole stick.

12

The Feminization of Society and the Poetic Dimension of Existence

In 1991, the great Argentine poet Roberto Juarroz introduced a new expression into the terminology of transdisciplinarity: the transdisciplinary attitude. Was it the privilege of a poet to be able to grasp one of the most important aspects of the transdisciplinary approach, simply by clarifying certain words?

The etymological meaning of the word *attitude* is "the aptitude for preserving a posture"; the opposite of posture is, of course, imposture.

From the transdisciplinary perspective, attitude is the individual or social capacity to preserve a constant, unchanging direction, no matter what the complexity of the situation or the hazards of life. On the social level, this direction is that of the flow of information crossing the different levels of Reality, whereas on the individual level, it is that of the flow of consciousness crossing the different levels of perception.

To preserve a constant direction across the levels of Reality guarantees a growing effectivity (or effectivness:

ability to make something happen) of our action in the world and in collective life—that of a nation, of a people, of humanity in its entirety. The spectacular technological development, the pinnacle of which is the computer revolution, shows that this effectivity is actually present in History, notwithstanding the motivation of one or another of the actors in political, economic, or social life.

To preserve a constant direction across the levels of perception guarantees a growing affectivity (or feeling: influenced by or resulting from the emotions), which ensures the link between the whole and ourselves. Sages of all times have affirmed that knowledge of the self is an endless evolutionary process. From the beginning of humanity until the present, the great texts of mystics, of religion and literature, and of great works of art, all testify to the constant presence of affectivity in this world.

Harmony between Object and Subject presupposes harmony between the exterior space of effectivity and the interior space of affectivity. "Effectivity and affectivity" ought to be the slogan of a plan for civilization to meet the challenges of our time.

Alas, in today's world mere efficacy at any price exists, which is only a caricature of effectivity. Affectivity, on the other hand, has no commercial value: it is therefore scoffed at, ignored, forgotten, even scorned. In the final analysis, this contempt for affectivity is none other than contempt for the human being, who has been transformed into a commercial object. When there is a death of affectivity, there is necessarily a "death of man." This last expression has become very popular, and this is no historical fluke. Can one therefore be surprised at the dissolution of social norms; the degradation of social, political, and international ties; the growing violence in metropolitan centers; the refuge of the young within a cocoon of drugs and sects; or the massacres that are being constantly perpetrated on an Earth that paradoxically, today benefits from a human knowledge that is unprecedented? When a politician pronounces the word *love,* people look at him as if he were an extraterrestrial. The masters of this world, who hold in their (computerized) grasp all the wealth of the planet, do not feel threatened in the least by anything originating from the interior human space; such ideas are perceived by them as the remnants of a sweet, innocent utopia from another time. Yet it is the growing disequilibrium between effectivity and affectivity that puts our species in danger.

The threat of self-destruction of our species is not entirely negative because it has engendered its counterpart, which is self-birth. In my view, "the death of man" is after all a necessary stage in History, one that anticipates our second birth.

Taken together, all the levels of Reality and our knowledge of those levels designate what one can call the masculinity of our world. In turn, all the levels of perception and our knowledge of those levels designate the femininity of this world. Of course, the sex of human beings is not directly tied to either one sphere or the other. A man can very easily be found in the feminine sphere, or a woman in the masculine sphere.

As usual, it is a question of equilibrium, because a stick always has two ends. The face of the world is ternary: masculinity and femininity and the zone of maximal resistance between the levels of Reality and the levels of perception, that place where the wedding between the masculine and feminine can be celebrated.

An extraordinary, unexpected, and surprising Eros cuts through the levels of Reality and the levels of perception. Artists, poets, and mystics of all times have testified to the presence of this Eros in the world. Less known are the testimonies of great scientists attesting to the presence of this Eros in Nature. The joy of a great scientific discovery is of the same nature as the joy of a great artistic creation, and there is no question that mysterious paths of the imagination in both cases lead to these discoveries.

It is we who have wounded the Eros of the world by giving preference to the unbridled development of masculinity. Eros has been replaced by an erotic masquerade; the wedding of femininity and masculinity by a sexual liberation that has all the characteristics of slavery (to the extent that human beings become mere appendices to their own sex); and love has been replaced by attentive vigilance over the defense of territory. The inevitable consequence of the logic of the marketplace, or efficacy for efficacy's sake, is the social marginalization of women. The various feminist currents that have cut through the twentieth century testify to this constant marginalization. But feminism, in turn, can find a much more solid basis for reflection and action than it has at present, in the necessary equilibrium between masculinity and femininity.

All future projects of civilization must necessarily pass through a process of social feminization. Just as woman, not man, gives birth to a child, it is the feminization of our world that can give birth to the social ties that are now so conspicuously absent in the cross-communications of human beings.

This process in no way implies a social, political, cultural, philosophical, or religious homogenization. By its very nature the transdisciplinary vision eliminates homogenization, which would reduce all the levels of Reality to one single level of Reality and reduce all the levels of perception to one single level of perception. The transdisciplinary approach presupposes a complex plurality and an open unity of the cultures, religions, and peoples of our Earth, and of the social and political visions in the heart of each and every people.

What can the definition of the transdisciplinary attitude be, in conformity with the transdisciplinary model of Reality?

When we locate ourselves on one specific level of Reality, we are fatally caught within an endless chain of binary opposites: we are obliged to be for or against. The reconciliation on one and the same level of Reality between "for" and "against" is impossible: at most, one arrives at a compromise that takes into account only part of the arguments for and part of those against, thus leaving both sides frustrated. Compromise can only be unstable: it is always a more or less long-term situation and it inevitably engenders a new pair of "for" and "against" opposites.

The reconciliation between "for" and "against" cannot be produced except by placing oneself on another level of Reality, where "for" and "against" appear as two contradictory poles of a larger unity, something that signifies being with, meaning that one takes into account everything that is positive and constructive within the "for" and the "against."

But, if one engages oneself exclusively in the crossing of different levels of Reality, this new behavior, to be with, neither for nor against, but both for and against, traps one in a new dogmatic, and even totalitarian system, even if through thought one changes levels of Reality. It is only through the harmony between levels of Reality and levels of perception, that is to say, through an accord between thought and one's own experience of life, that this trap may be avoided. Life is resistant to all dogma and all totalitarianism. The transdisciplinary attitude therefore presupposes both thought and interior experience, both science and consciousness; both effectivity and affectivity. The identity of arrows of the flow of information crossing the levels of Reality and of the flow of consciousness crossing the levels of perception assigns meaning or direction to the transdisciplinary attitude. The aptitude for preserving this posture orients one toward the compacting of information and consciousness that characterize the transdisciplinary attitude.

Thus, each thing and each being must find its proper place.

To be sure, we all look for one place: a place to live, a place to work and to satisfy the needs of existence, a place in the social hierarchy to satisfy the image we have of ourselves. But paradoxically, this place is practically never our own place, the place that would conform to the totality of our being. It is rare, very rare, when a human being on this Earth finds perfect harmony between his individual being and his social being.

This can show us the way of research for a true transpolitics: one founded on the inalienable right of all human beings to a harmonious interaction between their personal lives and their social lives. While each politician can and must remain true to his own political orientation, at the same time he must do everything possible to respect this inalienable human right. Transpolitics signifies neither the disappearance of politics nor the fusion of political approaches into one and the same "single thought." The complex plurality of the political order can accompany an open unity with a view toward satisfying one of the sacred rights of being human. The incommensurable riches of this Earth, the fabulous growth of knowledge, progressively powerful technological tools, and the treasures of wisdom and culture that have been accumulated since time immemorial all have the potential to transform that which could appear as a transdisciplinary utopia into an active reality.

However, in order to discover our own place in this world (one of the aspects of what is called "happiness") it is necessary to found some new, durable, and powerful social ties.

These new social ties can be discovered through research of the bridges between different areas of knowledge and between different beings, because the exterior space and the interior space are two facets of one and the same world. Transdisciplinarity can be understood as being both the science and the art of discovering these bridges.

It is there that we find the framework for an authentic revolution in intelligence. In itself, the explosive development of computer sciences does not equate with a revolution in intelligence. In the absence of affectivity, the effectivity of computers proves a dry, dead, even dangerous road, one more challenge of modernity. Intelligence is the capacity to read between the lines of the book of Nature and between the lines of the book of the interior being. In the absence of bridges between beings and things, technoscientific advances function only to intensify an increasingly incomprehensible complexity.

What is a dialogue between two beings in the absence of the bridge of a common language? Two parallel discourses engendering endless misunderstandings. What is social dialogue in the absence of bridges between social partners? A market of fools that only aggravates the social fracture. What is a dialogue between nations, states, and peoples of this Earth in the absence of bridges between them? A temporary adjournment of the final confrontation. A true dialogue can only be transdisciplinary, founded on the bridges that link beings and things at the deepest level.

The quantum revolution and the computer revolution are of no help in our daily life if they are not accompanied by a revolution in intelligence. It is thus that the wedding between the femininity and the masculinity of the world can be celebrated. "It is engagement in modern life which will make our existence a revolutionary act of creation," writes Jean Carteret.

The word *revolution* is not devoid of meaning because of the defeat of the social revolution. Today, revolution can only be a revolution in intelligence, which transforms our individual and social life into an aesthetic as well as an ethical act, an act that unveils the poetic dimension of existence. In our time, an effective political will can only be an affective poetic will. This can appear as a paradoxical and provocative proposition in a world animated by the exclusive concern with efficacy for efficacy's sake, where competition is pitiless, where violent confrontation is permanent, and where the number of those excluded from the feasts of consumerism and knowledge will continue to grow. By excluding more and more, we will end by excluding our own existence on the surface of this Earth.

The word *poetic* comes from the Greek word *poiein,* which means "to do": To do, today, means to reconcile the contradictories, to reunify the masculinity and femininity of the world.

13

On the Cult of Personality

The most obvious and extreme manifestation of the masculinity of our world is the appearance of the cult of personality at every level of social life.

Classic thought bequeaths to us a burdensome heritage: the dogma of the existence of a single level of Reality. In the absence of any vertical dimension, inevitably the image becomes more important than Reality, and fantasy creeps in between our eyes and Reality.

Still worse, today Reality must conform to the image we make of it. The televised images that invade our homes daily abundantly illustrate this assertion. A head of state becomes ill on a live broadcast, and transmission is immediately suspended, because this image does not conform to the authority of a head of state. If a beautiful actress becomes old or sick, her image is no longer shown.

The mask becomes more important than the face. There is only one face, but multiple masks. While functioning as a necessity for individual or social life, the mask—the persona —corresponds to a certain personality. Continual disharmony between individual and social life produces multiple personalities in one and the same being. The contradictions and conflicts between different personalities inside a person lead to the dissolution of interior being, which is no longer recognizable under these multiple masks. In these conditions how can one imagine a viable social connection? When one person speaks to another, how can we know which mask is speaking?

We now live by proxy. We delegate our lives to some authority figure, to some guru, to the image of some singer or athlete. Madonna is better known today than the Virgin Mary. And does anyone complain?

We can even affirm that this multiplicity of personalities is the basis of a consumer society. In general, the growth of consumerism is calculated to be a function of the number of persons who are susceptible to consuming. However, any given person may exhibit multiple personalities, so the number of potential consumers is much larger than the number of persons who consume. Advertisers have long understood this relatively trivial evidence, which, like all evidence, is not very visible. Every day these advertisers stimulate a different desire, and each desire creates a new potential consumer in one and the same person. A human being's material needs are limited, but his desires are unlimited. The consumer society has a beautiful future before it throughout the world. It is of little importance that the more one consumes, the less one is. What is important is consumption, even if this leads to the consumption of being. Can communion between beings be based on consumption?

Of course, we understand the extreme and monstrous forms of the cult of personality much better in the form of dictators, both great and small. These extreme forms exhibit the very essence of the phenomenon of the cult of personality: the confusion of place. How could a man destined to be a painter become the dictator of a great people and coldly exterminate a haunting number of human beings? How could another man destined to be a village priest become the dictator of a great country, imprisoning and exterminating millions of human beings in gulags? Those two tyrants who bloodied the Earth should have been able to remain in their places, that of a painter or a village priest, and thus might have ended their days quite happily. How can an empty shell be inhabited by an infinite number of fantasies? How can a man believe he has become the God of a people? The break between the interior space and exterior space of a human being can bring an interesting clarification to this process: When interior space is reduced to nothing, exterior space can become monstrous.

Each being has a place and can be happy if he remains in that place. There is no place that is more degrading than any other, no place more enviable than any other. The only place that invites us is our own place, and it is unique, to the extent that we are each unique. Yet, to find our proper place, by conforming our interior being to our exterior being, is an extremely difficult process, one that a society founded on effectivity makes practically impossible. We always want the place of another.

Our only authority is that of our interior experience and our work. It is of little importance whether this work is anonymous or celebrated. The greatest work—the Great Work—is our own life.

The greatest cathedrals were built over several centuries. Most of the names of the builders of cathedrals are still unknown to us. But the work is there, illuminating the life of small and large cities.

A multidimensional, multireferential Reality is incompatible with the cult of personality. Multiple masks fall in order to yield the place to the vertical face of being. A new sense of equality between human beings is gradually designed: the inalienable right of each being to find his own place. A man becomes free only when he finds his proper place. Human brotherhood consists in helping each other find that place.

For that, humanity is obliged to build its own body. It is an ensemble of subjects who build the Subject; it is an ensemble of human beings which build the Human. In a body, each cell has its place. A viable society passes through a polyphonic accord between subjects, between their different levels of perception and their different levels of knowledge.

One day perhaps humanity will be at once a complex plurality and a complex unity. Perhaps. If we truly want it. More precisely: if the secretly included middle in us wants it.

14

Science and Culture: Beyond Two Cultures

At the beginning of human history science and culture were inseparable. They were animated by the same questions, those about the meaning of the universe and the meaning of life.

In the Renaissance that tie was not yet broken. As its name indicates, the first university was dedicated to studying the universal. The universal was embodied in those who would make their stamp on the history of knowledge. Cardano, the inventor of imaginary numbers and of the suspension system that bears his name, was a mathematician, a doctor, and an astrologer; the same person who established the horoscope of Christ was the author of the first systematic exposition of the calculus of probabilities. Kepler was both an astronomer and an astrologer. Newton was simultaneously a physicist, a theologian, and an alchemist. He was as captivated by the Trinity as by geometry, and he spent more time in his alchemical laboratory than in the elaboration of his *Philosophiae Naturalis Principia Mathematica*. The founders of modern science had nothing in common with the stereotypical image of a scientist. In our day, Reality must conform to image in this sphere as well. Paradoxically, the scientist is forced, in spite of himself, to become a high priest of truth, an embodiment of rigor and objectivity. The complexity of the birth of modern science and of modernity itself helps us to understand the subsequent complexity of our own time.

The germ of the split between science and meaning, between subject and object, was certainly present in the seventeenth century, when the methodology of modern science was formulated, but it did not become full-blown until the nineteenth century, when the big bang hypothesis took flight.

In our time, the split was consummated. Science and culture have nothing more in common; this is why one speaks of science and culture. Every self-respecting government has a minister of culture and a minister of science. Every self-respecting international institution of higher education has a department of culture and a department of sciences. Those who try to cross the frontiers discover the risks of such an adventure. Science does not have access to the nobility of culture, and culture does not have access to the prestige of science.

Within science one distinguishes the exact sciences from the human sciences, as if the exact sciences were inhuman (or subhuman) and the human sciences inexact (or nonexact). Anglo-Saxon terminology is still worse: one speaks of hard sciences and soft sciences. We will pass over the sexual connotation of these terms, in order to explore their meaning. What is at stake are the ideas of definition, rigor, and objectivity, which convey the sense of exactitude (or of "hardness"). At base, according to classical thought, the only exact definition is a mathematical definition, the only rigor worthy of its name is mathematical rigor, and the only objectivity is that corresponding to a rigorous mathematical formalism. The "softness" of the human sciences attests to their lack of respect for these three key ideas, which formed a paradigm of simplicity over the course of several centuries. What could be softer, more complex, than the subject himself? The exclusion of the subject is therefore a logical consequence. "The death of man" coincides with the complete separation between science and culture.

One understands the indignant cries unleashed by the concept of two cultures—scientific and humanist culture—introduced some decades ago by C. P. Snow, a novelist and a scientist. The emperor wore no clothes. The comfort of the owners of the spheres of knowledge was threatened and their conscience was put to the test. Science is certainly part of culture, but this scientific culture is completely separated from humanist culture. The two cultures are perceived as antagonists. The split between the two cultures is first of all a split between values. The values of scientists are not the same as the values of humanists. Each world—the scientific world and the humanist world—is hermetically shut on itself.

The debate created by the concept of "two cultures" has been beneficial, because it has imparted a sense of danger to their split: it has exposed the extreme masculinization of our world, with all the dangers that brings to our individual and social life.

In recent times the signs of a reconciliation between the two cultures are multiplying, above all in the dialogue between science and art, the fundamental axis of a dialogue between scientific culture and humanist culture.

The attempts at rapprochement between art and science have had a primarily multidisciplinary character. Innumerable colloquia have reunited poets and astrophysicists or mathematicians, artists and physicists or biologists. Multidisciplinary initiatives have appeared in secondary and university-level teaching. These attempts have the merit of revealing that dialogue between science and art is not only possible, but necessary.

One more stage has been passed through by the interdisciplinary rapprochement between science and art. Here too, the initiatives are numerous and fertile. The acceleration of this rapprochement has an unprecedented rhythm, which is produced before our eyes thanks to the information explosion: There is today a new kind of art that was born by transferring computer methods to the realm of art. The most spectacular example may be that of art that uses the incredible information circulating on the Internet as if it were new matter. Information rediscovers its original meaning of "in-formation": to create form, ceaselessly changing new forms, arising out of the collective imagination of artists. The interconnectivity of computer networks allows such connections between artists, who come together in real time on the Internet in order to create together, in song and image, a world that arises from somewhere else. This "somewhere else" is found in the inner worlds of artists trying to harmonize, to discover together whatever it is that connects them with creation. These experimental researches constitute the germ of a genuine transdisciplinarity in action.

It is here that the transdisciplinary method is shown to be indispensable, because all creation encounters a wall of representation. Regardless of the nearly unlimited power of computer networks, images created by several artists at once inevitably run counter to the limits of individual representation, which obviously varies from one artist to another. The juxtaposition of these different degrees of representation can only engender a chaotic virtual reality, devoid of order, notwithstanding its apparent beauty.

The encounter between different levels of Reality and different levels of perception engenders different levels of representation. Images corresponding to a certain level of representation have a different quality than images associated with another level of representation, because each quality is associated with a certain level of Reality and with a certain level of perception. Each level of representation appears like a veritable wall, apparently unassailable because of its relation to the images engendered by another level of representation. These levels of representation of the sensible world are therefore connected with the levels of perception of the creator, the scientist, or the artist. True artistic creation arises at the moment of crossing several levels of perception simultaneously, engendering a transperception. True scientific creation arises at the moment of crossing several levels of representation simultaneously, engendering a transrepresentation. Transperception permits a global, non-differentiated understanding of the totality of levels of perception. The surprising similarities between moments of scientific and artistic creation are thus explained, as brilliantly demonstrated by the great mathematician Jacques Hadamard.

In the example of computer art that was cited above, the practically unlimited informational power of computers allows a global simulation of the totality of levels of representation, through the mediation of mathematical language. Thus, for the first time in history, the human–computer interface, which has been masterfully explored by René Berger, potentially allows an encounter between transrepresentation and transperception. This surprising and unexpected encounter will certainly enable the future actualization of a potential creativity unsuspected in the human being. Thus, the transdisciplinary attitude is truly present there.

If multidisciplinarity and interdisciplinarity reinforce the dialogue between the two cultures, transdisciplinarity permits us to envisage their open unification. Besides the example of art and science, the preceding considerations on levels of Reality, perception, and representation, offer a methodological basis for the reconciliation of the two artificially antagonistic cultures: the scientific culture and the humanist culture, by virtue of their overlapping within the open unity of transdisciplinary culture.

15

The Transcultural and the Mirror of the Other

To contemplate twentieth-century culture is at once disconcerting, paradoxical, and fascinating. Since time immemorial, immense treasures of wisdom and knowledge have been accumulating, and still we continue to kill each other.

It is true that the treasures of one culture are virtually incommunicable to another. There are even more cultures than there are languages, and the number of languages on our planet is already legion. This is a formidable obstacle to authentic communication and communion between human beings, brought together by destiny on one and the same Earth. One can translate from one language to another, but translations are forced into merely apparent perfection at the cost of making more or less gross approximations. In the future one might imagine the appearance of a supercomputer, a kind of universal dictionary, capable of furnishing us with the translation of the words of one language into the words of another language. But a similar translation, be it partial or general, between different cultures is inconceivable, because cultures emerge from the silence between the words, and this silence cannot be translated. No matter what their emotional weight, the words of everyday life are primarily addressed to the intellect, the instrument granted to human beings for survival. But cultures emerge from the totality of human beings forming a collective in a particular geographic and historical area—all their feelings, hopes, beliefs, and questions.

Prodigious advances in methods of transportation and communication have brought about an intermingling of cultures. Today one finds more Buddhists in California than in Tibet and more computers in Japan than in France. This cultural intermingling is chaotic. The proof: the innumerable difficulties that are concomitant with the "integration" of different cultural minorities into various countries in the world. Under what banner can this phantasmagoric integration be wrought? No Esperanto, no Volapük, no matter how computerized it might be, can ever promise a translation between cultures. Paradoxically, today everything is open and closed at the same time.

The overwhelming advance of technoscience has served only to deepen the abyss between cultures. The nineteenth-century hope for a single culture in a worldwide society, founded on the happiness brought about by science, crumbled a long time ago. In its stead, we have witnessed a complete separation between science and culture on the one hand, and, on the other, a cultural fragmentation within each and every culture.

The separation between science and culture engendered the myth of a separation between the West and the East: the West as the repository of science and knowledge of Nature, and the East as the repository of wisdom and knowledge of the human being. This separation, at once geographic and spiritual, is artificial, because, as Henry Corbin has stated, the Orient is a facet of the

Occident and the Occident is a facet of the Orient. In each human being the Orient of wisdom (the affective) and the Occident of science (the effective) are potentially reunited. But like all myths, that of the separation of the wisdom of the East and the science of the West is partly true, because modern science really was born in the West. And, in fact, it is true that the spread of the Western lifestyle throughout our planet is associated with the destabilization of traditional cultures. On account of its economic strength, the West has a great responsibility: to avoid the cultural disintegration resulting from the unbridled development of technoscience.

Cultural fragmentation is felt in the heart of every culture. The disciplinary big bang has its equivalent in the big bang of cultural modes. As the inevitable result of the loss of frames of reference in an increasingly complex world, one mode of thought is swept away by the next with ever increasing speed. Before long, through the intervention of computers, the speed of change in cultural modes may reach the speed of light. But although, thanks to scientific methodology, the disciplinary fragmentation within science leads to more or less stable territories, those of cultural modes remain the domain of ephemera. Culture today appears more and more like some kind of monstrous rolling garbage can in which strange defenses against the terror of nonmeaning proliferate. Of course, as always, the new is hidden in the old, yet it is slowly but surely being born. This still formless mixture of the new with the old is fascinating, because beyond all the different cultural modes, a new cultural way of being is taking shape.

In spite of its chaotic appearance, modernity leads to a rapprochement between cultures. With infinitely more intensity than in previous times, modernity brings about a resurgence of the need to unite being with the world.

The potential for the birth of a culture of hope is precisely equivalent to the potential for self-destruction that is engendered by the abyss of non-meaning.

The multicultural shows that the dialogue between different cultures is enriching, even if its goal is not real communication between cultures. The study of Chinese civilization was certainly fruitful for deepening comprehension of European culture. The multicultural helps us discover the face of our own culture in the mirror of another culture.

The intercultural is clearly assisted by the growth of transportation and communication and by economic globalization. Deepening discovery of hitherto badly known or unknown cultures makes unsuspected potentialities burst forth from our own culture. The influence of African art contributed to the appearance of Cubism—an eloquent example. The face of the Other permits us to know our own face better.

Obviously, the multicultural and the intercultural by themselves do not assure the kind of communication between all cultures that presupposes a universal language founded on shared values, but they certainly constitute important steps toward the act of transcultural communication.

The transcultural designates the opening of all cultures to that which crosses through them and transcends them.

The reality of an opening such as this is proven, for example, by the research that has been led for a quarter of a century by director Peter Brook with his company, Centre International de Créations Théâtrales. The actors are of different nationalities and thus are themselves immersed in different cultures. Nevertheless, during a performance they reveal qualities that cross and transcend cultures, using a wide range of material from the

Mahabharata to *The Tempest*, from *The Conference of the Birds* to *Carmen*. The popular success of these performances in different countries of the world shows us that such a transcultural approach can be as accessible to audiences as their own culture.

The perception of that which crosses and transcends cultures is first of all an experience that cannot be reduced to the merely theoretical, yet it is rich with teachings for our own life and for our action in the world. It indicates that no one culture constitutes a privileged place from which one can judge other cultures. Each culture is the actualization of a potentiality of the human being, in a specific place on Earth and at a specific moment in history. Different places on Earth and different moments in history actualize different potentialities of the human being, that is, different cultures. It is the open totality of the human being that constitutes the "place without place" of that which crosses and transcends cultures.

The perception of the transcultural is first of all an experience, because it concerns the silence of different actualizations. The space between the levels of perception and the levels of Reality is the space of this silence; it is the equivalent, in interior space, of that which is called the quantum vacuum in exterior space. It is a full silence, structured in levels. There are as many levels of silence as there are correlations between levels of perception and levels of Reality. And beyond all these levels of silence there is another quality of silence, that place without place which the poet and philosopher Michel Camus calls "our luminous ignorance." This nucleus of silence appears to us as an unknowable, because it is the unfathomable well of knowledge, but this unknowable is luminous because it illumines the very structure of knowledge. The levels of silence and the levels of our luminous ignorance determine our lucidity. If there is a universal language, it goes beyond words, because it concerns the silence between the words and the unfathomable silence that is expressed by each word. Universal language is not a language that can be captured in a dictionary; it is the experience of the totality of our being, reunited at last, beyond all its myriad forms. It is, by its very nature, a translanguage.

From the physical perspective, human beings are the same: they are constituted by the same matter, above and beyond their various physical structures. Human beings are the same from the biological point of view: the same genes engender different skin colors, different facial expressions, our qualities, and our faults. The transcultural suggests that human beings are also the same from the spiritual point of view, beyond the enormous differences that exist between various cultures. The transcultural is expressed by simultaneously reading all of the levels of silence, across a multitude of cultures. "The rest is silence," as in the last words of *Hamlet*.

It is the Subject who forges translanguage, an organic language, which captures the spontaneity of the world, beyond the infernal chain of abstraction after abstraction. The event of being is just as spontaneous and sudden as a quantum event. It is the sequence of events of being that constitutes true actuality that, alas, does not receive any attention from our mass media. And, yet, these events are what constitute the nucleus of true communication.

Finally, that which is found in the center of the transcultural is the problem of time. Time is the measure of change of different processes. As a result, time is always thought of in the past or in the future. It is the domain of the Object. In contrast, time lived in the spontaneity of an event of being, the present instant, is unthinkable. As Charles Sanders Peirce, one of the great precursors of transdisciplinarity, observed, the present moment is a point in time in which no thought can occur, no detail can be separated.

The present moment is living time. It concerns the Subject; more precisely it concerns that which links the Subject to the Object. The present instant is, strictly speaking, a non-time, an experience of relation between Subject and Object; thus, it contains potentially within itself the past and the future, the total flow of information that crosses the levels of Reality and the total flow of consciousness that crosses the levels of perception. The present time is truly the origin of the future and the origin of the past. Different cultures, present and future, have extension in the time of history, which is the time of change in the state of being of peoples and of nations. The transcultural concerns the time present in transhistory, which concerns the unthinkable, the unthought, and epiphanic.

The transcultural is the necessary condition for the existence of culture. Michel Cazenave conceives it under a double rubric: as the differentiating unity of the cultures, defining what it means to be human, and the neverending circulation between cultures, functioning to preserve them from disintegration. In effect, the complex plurality of cultures and the open unity of the transcultural coexist in the transdisciplinary vision. The transcultural is the spearhead of transdisciplinary culture. Different cultures are the different facets of the human being. The multicultural allows the interpretation of one culture by another culture, the intercultural permits the fertilization of one culture by another, and the transcultural assures the translation of one culture into various other cultures, by deciphering meaning that links them, while simultaneously going beyond them.

Transcultural language, which both enables the dialogue between all cultures and prevents their homogenization, is one of the major aspects of transdisciplinary research.

16

Transdisciplinarity—Deviations and Wrong Turns

Great changes in history and culture often have been brought about by a minute deviation: a small divergence in respect to the governing norms suddenly unleashes the unraveling of the system in place and later allows the emergence of new all-powerful norms.

In the realm of history the most striking example is probably that of the birth of Christianity. Armed only with their vision of another world, some illuminés initiated a movement that was to change the world.

In the scientific realm, small anomalies on the experimental level were the source of the two great intellectual formulations of this century—the theory of relativity and quantum mechanics. It was not possible to eliminate these anomalies in spite of considerable theoretical efforts. Such anomalies have thus generated an unprecedented expansion of the realm of scientific truth. The new norms have unequivocally ruled twentieth-century physics.

An all-powerful social or cultural system is therefore often only a successful deviation. Of course, the mere fact of being a deviation is not sufficient for success. So what does make for a successful deviation?

An analysis of the parameters that must be taken into account in order for a deviation to be successful quickly leads to an impasse, because for the most part we do not know the number or even the kind of these parameters. In the language of a physicist one can affirm that in the case of a deviance, initial conditions are less important than the nature of the laws operative in the domain being considered. A deviation that succeeds is in conformity with that which is most central to these laws, with its central tendency. The deviance acts as a vision opening toward a level of Reality different from the level at which the system under consideration is situated. The Gödelian structure of Nature and knowledge relates directly to the success of a deviation.

By nature transdisciplinarity has the status of deviation, not of dissidence (which always ends by being absorbed by the system in place). Transdisciplinarity diverges from the supposedly indisputable norm of efficacy where there are no breaks and no other values than efficacy itself—a norm that seems to be founded on the proliferation of academic and nonacademic disciplines. Transdisciplinarity acts in the name of a vision of the equilibrium necessary between the interior and the exterior of the human being—a vision connected to a new Reality different from that of the actual world. Must one conclude, however, that transdisciplinarity is a deviation that is going to succeed? We leave the pain of responding to this question to those who will live at the time of the next millennium. For the time being we must clear away some of the major obstacles to transdisciplinarity—obstacles that can be qualified not just as deviations, but as wrong turns.

In the case of transdisciplinarity, such wrong turns are rigorously defined. They are generated by levels of confusion, a relevant transdisciplinary idea introduced by Philippe Quéau.

Levels of confusion are generated by lack of respect for the unique and singular role that each level of Reality and each level of perception plays within the open unity of the world. Thus, these wrong turns are legion. But one can nevertheless designate those that threaten transdisciplinarity by a more or less deceitful reduction of it to something that it is not—thus to eliminate deviance by a return to the operative norms, in the very name of this deviation.

This wrong turn is particularly insidious since it attempts to eliminate the deviation (i.e., in this case, transdisciplinarity) by a tacit return—carried out in the name of the deviation itself—to the norms in force.

The most elementary confusion consists in forgetting the discontinuity of levels of Reality and levels of perception, replacing them implicitly with continuity. Then an inevitable reduction of all levels of Reality and all levels of perception to one and the same level occurs; complex plurality is reduced to a complexity without any order other than the horizontal one of levels of organization, and the open unity of the world becomes a plural world closed in upon itself, and thus favorable to all ideological and dogmatic mixups. This zero level of confusion is therefore very dangerous. It implies the confusion between multidisciplinarity, interdisciplinarity, and transdisciplinarity. Harmonious dialogue between disciplinarity, pluridisciplinarity, interdisciplinarity, and transdisciplinarity, where each complements the others, would thus be replaced by the cacophony of endless, uninteresting semantic games.

But there are some other wrong turns in transdisciplinarity, more subtle, and consequently more dangerous.

Two extreme levels of confusion are possible.

One can envisage the arbitrary reduction of all the levels of perception to one and the same level of perception even while recognizing the existence of several levels of Reality.

This level of confusion can lead to a new scientism being taken as the intellectual foundation of a badly understood transdisciplinarity. The scientistic position is based on the belief that one type of knowledge—Science—has the only access to truth and reality. The scientistic ideology of the nineteenth century proclaimed that only science could lead to the discovery of truth and reality.

The neoscientism breeding today no longer denies an interest in a dialogue between science and other areas of knowledge, but, for all that, it still holds fast to the postulate of the limitless horizon and the pertinence of science and that science remains capable of totally accounting for everything that exists.

The most characteristic sign of neoscientism is the negation of the value of all research dealing with metadiscourse or metatheory. Thus, everything becomes just a (potentially deadly) game or (potentially destructive) child's play: the human being amusing himself by leaping from one branch of knowledge to the other, but never finding any bridge that links one way of knowing to the other.

The same level of confusion can lead to the absorption (and therefore the destruction) of transdisciplinarity by extremist ideologies from either extreme, from the Right or from the Left, looking for renewed virginity.

We live in a troubled world where anything can happen. The void created by the unexpected, warless implosion of the Soviet empire will be quickly filled because History, like Nature, has a horror of a vacuum. Slogans such as "the end of History" or "the death of ideologies" try to hide this vacuum, which will soon be replaced, for better or worse. In our day extremists dare not present themselves as extremists, because they know that their chance of success is practically nil. Then the wolf will wear sheep's clothing, thanks to neoscientistic ideology. Can one imagine what a Hitler or a Stalin in our era might be like—armed with the power of computers and the ability to manipulate genes, able to juggle the spiritual account books of contemporary human beings? The recognition of the existence of several levels of Reality could lead to a false sense of freedom and spirituality, justifying every manipulation imaginable.

Neoscientism and extremist ideologies have in common an obsession with research into the death of the Subject. Interior man is the nightmare of all scientism and of all totalitarian dogma, whatever their disguise.

Another extreme level of confusion consists in the recognition of the existence of several levels of perception, all the while refusing to admit the existence of several levels of Reality.

This wrong turn leads to the annexation of transdisciplinarity by hermetic irrationalism, which presently is experiencing a rather inevitable resurgence (is not irrationalism the twin brother of extreme rationalism?). Transdisciplinarity would thus be quickly emptied of all life by being transformed into a pure phenomenon of language, a language for the "initiated": one could thus speak "transdisciplinary" as one might speak "Lacanian" (this last assertion obviously is not intended to make any unseemly reference to Lacan himself): a language that would say everything about nothing. Two strong tendencies with no apparent link between them could lead to this wrong turn. On the one hand, there is the current infatuation for cheap esotericism: one keeps the language of alchemy, but one forgets that it connects with precise interior experiences; one keeps the language of astrology, but otherwise forgets that its symbols were linked to a science of psychological types, and so forth. On the other hand, the current academic style is to reduce everything to language: as if there were no Reality, in the ontological sense of the term, but merely languages that construct a reality, and no science which explores Nature, but merely a social construction that we call "science."

In actual fact, these two tendencies explain the drifting off course of the present society; indeed, they cloak themselves with the attractive embellishments of spirituality or of academic respectability in order to discreetly hide this wrong turn.

Thus, there is a new level of confusion, halfway between the zero level of confusion and the extreme levels of confusion. One can easily recognize the existence of several levels of Reality and of several levels of perception without nevertheless taking into consideration their rigorous correlation.

In this context, the most obvious wrong turn would consist of the assimilation of transdisciplinary enthusiasm by the New Age. Here it is not a question of making a value judgment on all the components lumped together as the "New Age," where one finds both the best and the worst. This complex, chaotic, and anarchic movement, would demand a nuanced judgment, specific to the contradictory tendencies that comprise it. The source of the New Age is noble, because it tries to explain itself as a struggle for survival against the obsolescence and inadequacy of the current system of thought relative to the challenges of modern life. There is no doubt that some of the personalities who helped start the New Age were true innovators. After all, certain ideas and practices, above all those connected with the revaluation of the role of the body in the life of contemporary humanity, should not be dismissed. However, the danger associated with the New Age is rooted in its lack of rigor, which leads to unconsciously mixing everything into an amorphous catchall, where it will attempt to include transdisciplinarity as an honorable and more or less exotic component. Notwithstanding the motivations of one or another of its representatives, the New Age appears like a giant hypermarket of our consumer society, where each and every person can come to search for a little of the Orient and a little of the Occident, in order to regain, fairly cheaply, his peace of mind.

Spiritual consumerism is the mirror image of the consumption of material goods. The lack of rigor can lead to a sectarian insularity, with grave dangers. The rapid multiplication of sects is one of the signs of the disappearance of clear guidelines in a consumer society. The escape into the closed life of a sect actually indicates a need to avoid all responsibility in a world of incomprehensible complexity. The pseudo-spiritual drug is a drug like any other. Here as elsewhere it would be more intelligent to attack the causes of the malady instead of obsessively concentrating on the symptoms.

Another wrong turn of a similar nature would be the marketing of transdisciplinarity. Sold in such a way, transdisciplinarity could constitute an ideal means for bestowing new legitimacy to decision makers in distress without doing anything to change their approach. Can't one already see seminars for the business leaders in which Sufi spirituality walks hand in hand with the latest quantum physics, esotericism with neurophysiology, and Buddhism with the latest computer technology?

Certainly, this recent phenomenon of marketing is not wholly negative in itself. It is simply a question of opening the world of these decisions makers to the values of ancient or modern culture. But a danger indeed exists that what is most innovative in transdisciplinary culture is being co-opted more and more, sacrificing its true spirit to the god of efficacy for efficacy's sake, more so than ever before.

There is an urgent need for the formulation of a transdisciplinary deontology whose three major guidelines are recognizing (1) the inalienable rights of interior man, (2) the irreducible newness of our era, and (3) the atypical character of transdisciplinarity. This transdisciplinary deontology is one of the safeguards for the immutable orientation of the transdisciplinary attitude. This is why the participants in the First World Congress of Transdisciplinarity felt the need to elaborate a Charter (see the Appendix, page 147).

It is through separating transdisciplinarity from an acknowledgment of the rights of interior man (those which complement the rights of exterior man) that the worst distortions might be envisaged.

The acknowledgment of the irreducible newness of our era implies that today any return to an ideology, religion, or philosophy of the past is harmful. That does not exclude the riches of all the traditions of the world, but to the contrary, implies the rediscovery of them. It is the explicit acknowledgment of this newness that is one of the major guarantees of the absence of all such distortions. In transdisciplinarity, as in the quantum physics born at the beginning of the last century, one cannot make the new from the old.

The third major guideline for ensuring the absence of wrong turns is the acknowledgment of the atypical character of transdisciplinarity. The place of transdisciplinarity is a place without place. It is found neither in interior man (either generating a new religion, or a new philosophy, or a new metaphysics), nor in exterior man (generating a new science, which would be the science of sciences). One can thereby avoid the hollow but ever-active formulas such as "the death of man." The dialectical relation between history and trans-history demands that true transdisciplinary research be nourished in time and in history.

The transdisciplinary approach does not put holism and reductionism into opposition, but considers them as two aspects of one and the same knowledge of Reality. It integrates the local in the global and the global in the local. By acting on the local, one modifies the global and by acting on the global, one modifies the local. Holism and reductionism, or a global and local approach, are two aspects of one and the same multidimensional and multireferential world, the world of complex plurality and of open unity.

At its core, that which links all the wrong turns is the impoverishment of the trans-subjective dimension of being. Its distortion and its profanation risk enlarging the phenomena of irrationalism, obscurantism, and intolerance, of which the human, interhuman, and social consequences are intolerable.

By eliminating all the wrong turns, what appears is the long path that leads from knowledge to understanding in the name of rediscovered hope—a journey and endless quest that is begun anew.

17

Rigor, Opening, and Tolerance

Rigor, opening, and tolerance are the three fundamental characteristics of the transdisciplinary attitude.

Rigor is, first of all, the rigor of the language used in the reasoning based on the living knowledge—at once inner and outer—of transdisciplinarity.

Transdisciplinarity is both a body of thought and a lived experience. These two aspects are inseparable. Transdisciplinary language translates the simultaneity of these two aspects into words and actions. Any excessive slipping one way or the other—to the side of discursive thought or to the side of experience—takes us away from the domain of transdisciplinarity.

Transdisciplinary language is founded on the inclusion of the third force, that which is always found between the "why" and the "how," between the "who" and the "what." This inclusion is at once theoretical and experiential. A language exclusively oriented toward the "why," toward the "how," or toward even just the included middle does not belong to transdisciplinarity. The triple orientation of transdisciplinary language—toward the "why" and the "how" and toward the included middle—ensures the quality of presence of whoever or whatever uses transdisciplinary language. If I find the correct place in myself, at the moment in which I address myself to the Other, the Other can find the correct place in himself or herself and thus we can communicate. Because communication is first of all the correspondence between the right places in myself and in the Other, it is founded on authentic communion, beyond all dream or all fantasy or desire to manipulate the Other. Rigor is therefore also the research for the right place in myself and in the Other at the moment of communication.

This rigor is therefore the result of perpetual search, continually nourished by new knowledge and new experiences. The rigor of transdisciplinarity is of the same nature as scientific rigor but the languages are different. One can even assert that the rigor of transdisciplinarity is a deepening of scientific rigor to the extent that it takes into account not only things, but also beings and their relations to other beings and things. Taking account of all the givens present in a particular situation is a characteristic of this rigor. It is only in this way that rigor is truly a safeguard against all possible wrong turns.

Opening brings an acceptance of the unknown, the unexpected and the unpredictable.

There are three kinds of opening: the opening of one level of Reality toward another level of Reality, the opening of a new level of perception toward another level of perception, and the opening toward the zone of absolute resistance, which links the Subject and the Object. The unknown, the unexpected, and the unpredictable in a given moment in History is transformed, through time, into the known, the expected, and the predictable, while simultaneously giving birth to a new form of the unknown, the unexpected, and the unpredictable. The Gödelian structure of Nature and knowledge guarantees the permanent presence of the unknown, the unexpected, and the unpredictable. The source of their multiple forms in History is the zone of the absolute resistance that links Subject and Object. The opening of transdisciplinarity implies, by its very nature, the rejection of all dogma, all ideology, all closed systems of thought. This opening is the sign of the birth of a new type of thought turned not so much toward answers as questions. The Subject is himself the unfathomable question that assures the permanence of questioning. The rejection of questioning, absolute certitude, is the mark of an attitude beyond the realm of transdisciplinarity. Transdisciplinary culture is a culture of questioning, one perpetually accompanied by responses accepted as temporary.

Tolerance results from accepting the fact that ideas contrary to the fundamental principles of transdisciplinarity exist.

The transdisciplinary model of Reality clarifies the old problem of tolerance in a new way. Harmony between the levels of Reality and the levels of perception can be strengthened or weakened over time, either evolving or involving. There is therefore a problem of choice. Transdisciplinarity resolutely makes an evolutional choice, but we are forced to accept the existence of the opposite choice as well. The involutional choice implies the increase of binary oppositions or antagonisms, but the role of transdisciplinarity is not to struggle against this involutional choice, for an opposing choice is also inscribed in the nature of the Subject, and the struggle would end up, eventually, reinforcing the involutional choice, since the levels of action of the transdisciplinarity and anti-transdisciplinarity are different. The role of transdisciplinarity is to work in the direction of its own choice, and to show through action that what surpasses binary oppositions and antagonisms can be realized effectively.

Rigor, opening, and tolerance must be present in transdisciplinary research and practice.

The field of research and the practice of transdisciplinarity is immense, ranging from the fertilization of disciplinary research to the elaboration of a master plan for civilization. In this context, it is useful to introduce the concept of degrees of transdisciplinarity.

The degrees of transdisciplinarity are defined by taking into account, more or less completely, the three methodological pillars of transdisciplinarity: the levels of Reality, the logic of the included middle, and complexity.

The first degree of transdisciplinarity concerns the disciplines themselves. It is the spirit of a researcher in such and such a discipline that can be transdisciplinary. All disciplines can be animated by the transdisciplinary attitude: there is no single discipline that is favored over another from the point of view of transdisciplinarity. There are degrees of transdisciplinarity, but one cannot have favored disciplines with a transdisciplinary character.

From all the evidence, the transdisciplinary method does not replace the methodology of each discipline, which remains as it is. Instead the transdisciplinary method enriches each of these disciplines, by bringing them new and indispensable insights, which cannot be produced by disciplinary methods. The transdisciplinary method could even lead to some true discoveries at the heart of disciplines. This is natural, because one aspect of transdisciplinarity is research that crosses disciplines. An extremely eloquent historical precedent is the example of Oersted who, while exploring an idea from *Naturphilosophie*—that of polarity—was led to the scientific discovery of electromagnetism. A contemporary example might be, in the near future, the emerging field of quantum computers, which will unify fundamental science (quantum mechanics) and high technology: the quantum computers might unexpectedly bring the included middle into the heart of everyday reality.

Similarly, transdisciplinarity can fertilize multidisciplinary and interdisciplinary research, by opening them toward the common space of Subject and Object.

The penetration of the transdisciplinary attitude into the fields of poetry, art, esthetics, religion, philosophy, and social sciences is of very particular interest. Within each of them, another degree of transdisciplinarity becomes active, which implies not only that which crosses the disciplines but also that which structures them. At the base of each discipline there is a transdisciplinary attitude that gives the discipline its meaning; because, in the depths of every discipline is found the abyss that links the transdisciplinary Subject and Object.

18

The Transreligious Attitude and the Presence of the Sacred

The problem of the sacred, understood as the presence of something irreducibly real in the world, is unavoidable for any rational approach to knowledge. One can deny or affirm the presence of the sacred in the world and in ourselves, but, in view of elaborating a coherent discourse on Reality, one is always obliged to refer to it.

The sacred is that which connects. The sacred links, as indicated by the etymological root of the word *religion* (*religare*—"to bind together again"), but such an ability is not, in and of itself, an attribute of just one religion: Mircea Eliade once stated in an interview: "The sacred does not imply belief in God, in gods, or spirits. It is . . . the experience of a reality and the source of the consciousness of existing in the world." The sacred is first of all an experience; it is transmitted by a feeling—the "religious" feeling—of that which links beings and things and, in consequence, induces in the very depths of the human being an absolute respect for the others, to whom he is linked by their all sharing a common life on one and the same Earth.

The abolition of the sacred led to the abomination of Auschwitz and to twenty-five million deaths under the Stalinist system. The absolute respect for others has been replaced by the pseudosacralization of a race or of a new man, embodied by dictators elevated to the rank of divinities.

The origin of totalitarianism is found in the abolition of the sacred. While it is the experience of the irreducibly real, the sacred is actually the essential element in the structure of consciousness and not simply a stage in the history of consciousness. When this element is violated, disfigured, mutilated, history becomes criminal. In this context, the etymology of the word *sacred* is highly instructive. This word comes from the Latin *sacer,* which is to say, "that which cannot be touched without soiling," but also, "that which cannot be touched without being soiled." *Sacer* designates the guilty, who are consecrated to the infernal gods. At the same time, because of its Indo-European root *sak,* sacred is linked to *sanctus,* "holy." This double meaning of *sacer*—sacred and evil—is the double meaning of History itself, with its stammerings, its contortions, and its contradictions that give the impression that History is a tale of madmen.

In 1955, André Malraux was quoted by a French newspaper as saying, "With psychoanalysis, our century rediscovered the demons in man—the endeavor which awaits us now is rediscovering the gods." It is paradoxical and significant that the most desacralized period in History has generated one of the most profound reflections on the question of the sacred. The unavoidable problem of the sacred runs through the work of many diverse twentieth-century thinkers and authors, as well as that of scientifically minded artists and poets, from masters of thinking to masters of living.

The transdisciplinary model of Reality casts new light on the meaning of the sacred. A zone of absolute resistance links the Subject and the Object, the levels of Reality and the levels of perception. In the most general terms, movement is the simultaneous crossing of the levels of Reality and the levels of perception. This coherent movement is associated simultaneously with two meanings, two directions: an ascendant meaning (which corresponds to an "ascending" through the levels of Reality and of perception) and a descendant meaning (which corresponds to a "descending" through the levels). The zone of absolute resistance appears as the source of this double movement which is simultaneous and noncontradictory, ascending and descending through the levels of Reality and of perception. Absolute resistance is clearly incompatible with the attribution of just a single direction—of ascending or descending—precisely because it is absolute.

Relative to the levels of Reality and of perception, this zone is "beyond," but it is a "beyond" that is nevertheless connected to the levels. The zone of absolute resistance is the space of the coexistence of trans-ascendence and trans-descendence. As "trans-ascendence," this zone is bound up with the philosophical concept of transcendence (which comes from *transcendere*—"to climb beyond"). As "trans-descendence," it is bound up with the concept of immanence. Thus, the zone of absolute resistance is at once immanent transcendence and transcendent immanence, the former putting the accent on transcendence, whereas the second puts it on immanence. These two terms are therefore inadequate for designating the zone of absolute resistance, which appears as the irreducibly real that can neither be reduced to immanent transcendence nor to transcendent immanence. The word *sacred* is appropriate for designating this zone of absolute resistance, insofar as the included middle reconciles immanent transcendence and transcendent immanence. The sacred permits the encounter between the ascending movement and the descending movement of information and consciousness through the levels of Reality and the levels of perception. This encounter is the irreplaceable condition of our freedom and of our responsibility. In this sense, the sacred appears as the ultimate source of our values. It is the space of unity between time and non-time, causal and acausal.

There is an open unity of questioning in the multiplicity of answers because the sacred is the question.

One way or another, different religions, as well as agnostic and atheist currents, are defined in terms of the question of the sacred. Experience of the sacred is the source of a transreligious attitude. Transdisciplinarity is neither religious nor irreligious; it is transreligious. It is the transreligious attitude emerging from lived transdisciplinarity that permits us to learn to know and appreciate the specificity of religious and irreligious traditions that are foreign to us, to better perceive the common structures that found them, and thus, to arrive at a transreligious vision of the world.

The transreligious attitude is not in contradiction with any religious tradition or with any agnostic or atheistic current, to the extent that these traditions and currents recognize the presence of the sacred. In fact, the presence of the sacred is our transpresence in the world. If it were widespread, the transreligious attitude would make all religious wars impossible.

At its extreme point, the transcultural opens onto the transreligious. Through a curious historical coincidence, the discovery of the Venus of Lespugue occurred in 1922, just two years after the scandal of Brancusi's *Princess X*, a sculpture banned from the Salon des Indépendants due to accusations of obscenity. Astonished art lovers discovered the shocking resemblance between a paleolithic sculpture and that of the most innovative artist of the era, who would be later recognized as the founder of modern sculpture. Like the unknown sculptor of the Venus of Lespugue, Brancusi tried to make the invisible—the essence of movement—visible. Working in the context of their own cultures, each of these artists attempted to respond to the question of the sacred by making the invisible visible. In spite of the millennia separating the two creators, the forms that issued from their interior beings had a striking resemblance.

The transreligious attitude is not simply a utopian project—it is engraved in the very depths of our being. Through the transcultural, which leads to the transreligious, the war of cultures—an increasingly present menace in our time—has no more reason to be. If the transcultural and transreligious attitude were to find their proper place in modernity, the war of civilizations could not take place.

19

The Transdisciplinary Evolution of Education

The emergence of a transdisciplinary culture capable of contributing to the elimination of the tensions menacing life on our planet will be impossible without a new type of education that takes into account all the dimensions of the human being.

All the various tensions—economic, cultural, spiritual—are inevitably perpetuated and deepened by a system of education founded on the values of another century, and by a rapidly accelerating imbalance between contemporary social structures and the changes that are currently taking place in the contemporary world. More or less embryonic wars between economies, cultures, and civilizations never stop leading, here and there, to actual wars. In fact, our entire individual and social life is structured by education. Education is at the center of our becoming. The future is shaped by the education that is delivered in the present, here and now.

In spite of the enormous diversity of the systems of education that can be found in various countries, the globalization of the challenges of our era involves the globalization of the problems of education. The different upheavals continually confronting education in one or another country are only symptoms of one and the same flaw: the disharmony that exists between the values and the realities of a planetary life in the process of change. Most certainly, while there is no miraculous recipe, there is nevertheless a common center of questioning, which it would behoove us not to avoid if we truly want to live in a more harmonious world.

Growing awareness of a system of education that does not keep pace with the modern world is demonstrated by numerous recent conferences, reports, and studies. The latest and most exhaustive report was developed by the International Commission on Education for the Twenty-First Century chaired by Jacques Delors, in cooperation with UNESCO. The Delors Report strongly emphasized four pillars of a new kind of education: learning to know, learning to do, learning to live together, and learning to be.

In this context, the transdisciplinary approach can make an important contribution to the advent of this new type of education.

Learning to know means first of all training in the methods that help us distinguish what is real from what is illusory, and in the techniques that enable intelligent access to the fabulous knowledge of our age. In this context the scientific spirit, one of the highest goals ever attained in the human adventure, is indispensible. Precocious initiation into science is beneficial because it provides access, from the very beginning of human life, to the inexhaustible richness of the scientific spirit, which is based on questioning, and on the refusal of all a priori answers and all certitude contradictory to the facts. However, "the scientific spirit" does not at all mean the thoughtless increase of teaching scientific matters or retreating to an internal world based on abstraction and formalization. Although such excess is, unfortunately, still current, it can lead only to the exact opposite of the scientific spirit: previous ready-made answers are replaced by other new ready-made answers (this time, having a kind of "scientific" brilliance); thus, in the final analysis, one dogmatism is replaced by another. Nor is it the assimilation of an enormous quantity of scientific knowledge that gives access to the scientific spirit, but the quality of that what is taught. And here "quality" means to lead the child, the adolescent, or the adult into the very heart of the scientific approach, which is the permanent questioning related to the resistance of facts, images, representations, and formalizations.

Learning to know also means being capable of establishing bridges—between the different disciplines, and between these disciplines and meanings and our interior capacities. This transdisciplinary approach will be an indispensable complement to the disciplinary approach, because it will mean the emergence of continually connected beings who are able to adapt themselves to the changing exigencies of professional life, and who are endowed with a permanent flexibility that is always oriented toward the actualization of their interior abilities.

Learning to do certainly means acquiring a profession or a craft and the theoretical and practical knowledge that is associated with it. The acquisition of a profession or craft necessarily passes through a phase of specialization. One cannot do open-heart surgery if one has not learned surgery; one cannot solve a third-degree equation if one has not learned mathematics; one cannot be a producer without knowing theatrical techniques.

However, in our tumultuous world, in which the tremendous changes induced by the information revolution are but the portent of other still more tremendous changes to come, any life that is frozen into one and the same occupation can be dangerous, because it risks leading to unemployment, to exclusion, to a debilitating alienation. Excessive, precocious specialization should be outlawed in a world that is in rapid change. If one truly wants to reconcile the exigency of competition with the imperative of equal opportunity for all human beings, in the future every profession and every craft should be an authentically woven occupation, an occupation that would bind together in the interior of human beings threads joining them to other occupations. Of course, it is not simply a question of acquiring several competencies at the same time but of creating a flexible, interior core that could quickly provide access to another occupation, should that become necessary or desirable.

Here also, the transdisciplinary approach can be invaluable. In the last analysis, "learning to do" is an apprenticeship in creativity. "To make" also signifies discovering novelty, creating, bringing to light our creative potentialities. It is this aspect of "making" that is contrary to the boredom and sometimes even despair that is experienced by so many human beings who are obliged to exercise an occupation that does not conform to their interior predispositions simply in order to underwrite their basic needs. "Equal opportunity" also means the opportunity for the actualization of the creative potentialities, which vary from one person to the next. Competition could also mean the harmony of creative activities within a single community. Boredom and despair, the source of violence, conflict, and of moral and social resignation, can be replaced by the joy of personal realization, no matter what the place where this realization is effected, because a place can only be unique for each person at a given moment.

Creating the conditions for the emergence of authentic persons also means ensuring the conditions for the maximal actualization of their creative potentialities. The social hierarchy, so frequently arbitrary and artificial, could thus be replaced by the cooperation of levels structured in order to serve personal creativity. Rather than being levels imposed by a competition that does not take the interior being into account at all, these levels would in fact be levels of being. The transdisciplinary approach is based on the equilibrium between the exterior person and the interior person. Without this equilibrium, "to make" means nothing other than "to submit."

Of course, learning to live together first of all signifies respect for the norms that govern relationships between the beings comprising a collective. However, these norms must be truly understood and willingly internalized by each being, rather than obeyed out of submission to exterior constraints. "To live together" does not mean simply tolerating others' differences of opinion, skin color, and beliefs; submission to the exigencies of power; negotiating between the ins and outs of innumerable conflicts; definitively separating interior from exterior life; and merely appearing to hear the other while remaining convinced of the absolute rightness of our own position. If this is what it means, "living together" is inevitably transformed into its opposite: fighting each other. The transcultural, transreligious, transpolitical, and transnational attitude can be learned. To some extent, in each being there is a sacred, intangible core that is innate. Yet, if this innate attitude remains merely a potential, it can forever stay nonactualized, absent in life and in action. In order for the norms of a collective to be respected, they must be validated by the interior experience of each being.

There is one fundamental characteristic of the transdisciplinary evolution of education: to recognize oneself in the face of the Other. This is a question of permanent apprenticeship, which must begin in early childhood and continue throughout life. The transcultural, transreligious, transpolitical, and transnational attitude permits us to better understand our own culture, to better defend our national interests, to better respect our own religious or political convictions. Just as in all other areas of Nature and knowledge, open unity and complex plurality are not antagonists.

Learning to be appears at first like an insoluble enigma. We know how to exist, but how can we learn to be? We can begin by learning what the word *exist* means for us: discovering our conditioning, discovering the harmony or disharmony between our individual and social lives, testing the foundations of our convictions in order to discover that which is found underneath. In a building, the stage of excavation precedes that of foundation. In order to make a foundation for being, one must first of all proceed with the excavation of our certitudes, our beliefs, and our conditioning. To question, to question always; here also, the scientific spirit is a precious gift for us. This must be learned by the teachers as well as the taught.

"Learning to be" is also a permanent apprenticeship, in which teachers inform students as students inform teachers. The shaping of a person inevitably passes through a transpersonal dimension. The disrespect for this necessary process helps explain one of the fundamental tensions of our era, that between the material and the spiritual. The survival of our species largely depends on the elimination of this tension by means of a reconciliation between these two apparently antagonistic contradictions, which takes place on another level of experience than that of everyday life. "Learning to be" also means learning to know and to respect that which joins the Subject and Object. The other remains an object for me if I do not make this apprenticeship, which teaches me that together we, the Other and I, create the Subject joined with the Object.

There is one very obvious interrelationship between the four pillars of the new system of education: how to learn to make while learning to know, and how to learn to be while learning to live together?

In the transdisciplinary vision, there is also a transrelation which connects the four pillars of the new system of education and which has its source in our own constitution as human beings. This transrelation is like the roof that rests on four pillars of a building. If any one of the pillars of the building collapses, the entire building collapses, the roof with it. And, if there is no roof, the building falls into ruin.

A viable education can only be an integral education of the human being, according to the apt formulation of the poet René Daumal: an education that is addressed to the open totality of the human being, not to just one of the components.

At present, education privileges the intellect over the emotions or the body. This was certainly necessary in the previous era, in order to permit the explosion of knowledge. But this privileging, if it continues, sweeps us away in the mad logic of efficacy for efficacy's sake, which can only lead to our self-destruction.

Of course it is not a question of limiting or increasing the number of hours provided for artistic or athletic activities. This would be like trying to obtain a living tree by juxtaposing roots, trunk, branches, and leaves. This juxtaposition would only lead to a semblance of a living tree. Contemporary education concerns itself with only the leaves. But leaves do not make a whole tree.

Recent experiments made by the Nobel Prize–winning physicist Leon Lederman with children from disadvantaged neighborhoods of Chicago demonstrate what we have been saying. Lederman convinced some secondary school teachers to initiate new methods for teaching physics based on the touching of different objects and the discussion of contributions of different sense organs—sight, touch, smell—in the process. All this was play: it took place in an environment far removed from the usual formal apprenticeship in mathematics and physics. And the miracle happened. Even children who came from very poor families, where violence, lack of culture, and disinterest in the typical activities of children reigned supreme, discovered the abstract laws of physics through play. One year earlier, these same children had been declared incapable of ever understanding any abstraction. It is interesting, moreover, to point out that the greatest difficulties of the operation and—it goes without saying—the major part of its cost, were due to the resistance of the teachers: they had a great deal of trouble abandoning their old methods. The teaching of teachers proved longer and more difficult than the work with children.

The Chicago experiment shows very well that the intelligence assimilates knowledge much better and much more rapidly when this knowledge is also understood with the body and the emotions. In a living tree, the roots, the trunk, the branches and the leaves are inseparable: the sap, which ensures the life of the tree, moves vertically through all of them. This is the prototype of what was previously referred to as the "revolution of intelligence": the emergence of a new type of intelligence, founded on an equilibrium between analytic intelligence, feelings, and the body. It is only in this way that the society of the twenty-first century can reconcile effectivity and affectivity.

Transdisciplinarity education clarifies in a new way a need that is presently felt more and more—the need for a permanent education. In fact, transdisciplinary education, by its very nature, should take place not only in teaching institutions, from the kindergarten to the university, but also in the workplace—in fact, everywhere, and throughout our life.

In teaching institutions there is no need to create new departments and new chairs; this would be contrary to the transdisciplinary spirit. Transdisciplinarity is not a new discipline, and a transdisciplinary researcher is not some new kind of specialist. The solution would be to create work-groups for transdisciplinary research within every teaching institution. These work-groups would be the locus for gathering together a group of teachers and students from a particular institution who would generate and oversee their own organization and would all be animated by the transdisciplinary attitude. The same experiment could be carried out within various enterprises, and within other collectives, as well as within national and international institutions.

There is one particular problem that is posed by transdisciplinary education outside professional life. In a balanced society, the boundary between leisure time and apprenticeship time would gradually disappear. The information revolution could play a considerable role in our life for transforming training into pleasure and pleasure into training. The problem of unemployment of the young would certainly be alleviated by a hitherto unsuspected solution. In this context, grassroots efforts will play an important role in transdisciplinary education throughout life.

It is quite obvious that the various areas and ages of life call for extremely diverse transdisciplinary methods. Even if transdisciplinary education is a long-term, global process, it is still important to discover and to create places that help to initiate this process and ensure its development.

The university is the privileged place for an education geared toward the exigencies of our time which could also be the pivotal place for an education directed not only toward children and adolescents, but also toward adults.

In the transdisciplinary perspective, there is a direct and unavoidable relation between peace and transdisciplinarity. Severely fragmented thought is incompatible with the research of peace on this Earth. The emergence of a culture and an education of peace requires the transdisciplinary evolution of education and, especially, the transdisciplinary evolution of the university.

Instilling complex and transdisciplinary thought into the structures and programs of the university will permit its evolution toward its somewhat forgotten mission today—the study of the universal. The university could become a place of apprenticeship in the transcultural, transreligious, transpolitical, and transnational attitude, and in the dialogue between art and science, which is the axis of a reunification between scientific culture and artistic culture. A renewed university would become the place for welcoming a new kind of humanism.

20

Towards a New Humanism: Transhumanism

An expectant world. Expecting what? No clear-headed person can say for certain.

I do not know. All that I know is that our world is waiting. For whom? For what? For Woman perhaps, for Man also, and for their not yet celebrated union.

I do not know if "the madman" whom André Bourguignon describes so vividly will be able to face the challenges of the next century. Perhaps madness is the price we pay for our linguistic creativity, for our reason, for our genius. All I know is that if madness is the norm, then any wisdom that is opposed to the norm will also be a form of madness. In a world where everything is presented as a social construction; where the relative has become, in fact, the absolute; where violence is the other face of solidarity and exclusion is the other face of well-being; where the massacre of innocents is the other face of accord between peoples—it is unthinkable that we can find the true reason for living in such a world.

I do not know if there is a solution. All I know is that there is a "question": the question of the birth of an unknown, unpredictable world, wavering as it moves from closed space toward openness, toward the actualization of all possibilities. All we can do is bear witness. The present manifesto is just such a testimony.

Transdisciplinarity is not the way, but simply a way of bearing witness to our presence in the world and our lived experience linking the amazing ways of knowing in our time.

As Jacques Robin rightly emphasized, lived transdisciplinarity can lead us not only to a change in the ways we think but also in the ways that we behave. It is worthwhile examining the conditions that must be created in order to be able to bring about this new behavior.

From the point of view of transdisciplinarity, any closed system of thought, whatever it is, inevitably places the focus on the notion of a "mass," indistinct and unformed, an abstract concept that precludes giving any importance to the interior development of the human being. Nazi ideology places the focus on the mass that constitutes a "race," thus denying the innate nobility of all human beings, which led to the abominations of the concentration camps and gas ovens. In the name of noble ideas, communist ideology divinized the popular masses, constituted by identical "New Men," in contempt of the intrinsic heterogeneity of human beings, and this led to the crimes of the Stalinist era.

Liberal society is more correct and more balanced but it too places the focus on the concept of "mass"—privileging one or another social category or one or another profession. Certainly, while old ideals such as "liberté, égalité, fraternité" and "liberty and justice for all" are still proclaimed as sacred rights, society is still incapable of creating conditions for the effective realization of these alleged utopian values that would permit the reconciliation between exterior being (that which takes part in an apparently amorphous mass) and interior being (that which gives meaning to social life). A consumer does not equal a person, and it is the person who must be in the center of all civilized society. The only path to the reenchantment of the world is one that explores the infinite capacity of wonder within the human consciousness.

The implacable logic of efficacy for efficacy's sake can only serve the most frenzied egoism and, through individual or collective strategy, the profit of the rich and the detriment of the poor. Elephantiasis of the ego can never lead to the construction of a "person"; instead, it generates a conflicting coexistence of individuals engaged in a merciless competition, in the name of an efficiency whose rationality totally escapes even those who are its unconditional servants.

On the social level, the transdisciplinary vision, which is at once transcultural, transreligious, transnational, transhistorical, and transpolitical, leads to a radical change of perspective and attitude. Of course, no state should impose its structures upon the interior life of the human being, which must remain within the jurisdiction of total individual responsibility. But social structures should create conditions that will allow this responsibility to be realized and exercised. Economic development at any price must no longer be at the center of such social structures. Political economics and living are intimately connected. Creative research of a transdisciplinary political economics is founded on the postulate that it will serve the human being, not the reverse. Material well-being and spiritual well-being condition one another. I call the new form of humanism that offers each being the greatest capacity for cultural and spiritual development transhumanism. It involves searching for that which is between, across, and beyond human beings—that which could be called the Being of beings. Transhumanism is not aimed at a fatally destructive homogenization, but rather, at the maximal actualization of unity in diversity and diversity in unity. The focus will be placed not on the ideal organization of humanity through ideological formulas, which always bring about the contrary of whatever they advocate, but on a flexible structure oriented toward welcoming complexity. It does not involve defining the human being while searching to build "the New Man," for that approach always ends up destroying the human being by transforming a person into an object. And can an object have any kind of freedom besides that granted by the Grand Inquisitor in Dostoyevsky's *The Brothers Karamazov*?

Let us recall what has already been said: *Homo sui transcendentalis* is not a "New Man," but a person who has been born anew. *Homo sui transcendentalis* is the true natural state of the human being.

After all, what we find at the center of our questioning is the dignity of the human being, our infinite nobility. The dignity of the human being is also in the planetary and cosmic order. The appearance of the evolutionary human phenomenon on Earth is one of the stages of the history of the universe, just as the birth of the universe is one of the stages of human evolution.

The recognition of the Earth as a motherland is one of the imperatives of transdisciplinarity. Every human being has the right to nationality and, at the same time, to transnationality as well.

Transnationality in no way implies the devaluation or disappearance of nations. To the contrary, transnationality can only reinforce what is most creative and most essential in each nation. The word *nation* has the same root—*nasci*—as the word *Nature;* the noun form is *nationis,* which means birth. Nations can give birth to the transnational, and the transnational can eliminate national egoism, which leads to such murderous conflicts. The elephantiasis of nationalism has the same source as the elephantiasis of ego: nonrespect for the dignity of the human being.

When Pandora's box is opened, the evils that escape will threaten the human beings who populate the Earth. But hope and promise are also hidden in the bottom of the box. Transdisciplinarity bears witness to that hope and that promise.

Appendix I

The Charter of Transdisciplinarity

Preamble

Whereas, the present proliferation of academic and non-academic disciplines is leading to an exponential increase of knowledge which makes a global view of the human being impossible;

Whereas, only a form of intelligence capable of grasping the cosmic dimension of the present conflicts is able to confront the complexity of our world and the present challenge of the spiritual and material self-destruction of the human species;

Whereas, life on earth is seriously threatened by the triumph of a techno-science that obeys only the terrible logic of efficacy for efficacy's sake;

Whereas, the present rupture between increasingly quantitative knowledge and increasingly impoverished inner identity is leading to the rise of a new brand of obscurantism with incalculable social and personal consequences;

Whereas, an historically unprecedented growth of knowledge is increasing the inequality between those who have and those who do not, thus engendering increasing inequality within and between the different nations of our planet;

Whereas, at the same time, hope is the counterpart of all the afore-mentioned challenges, a hope that this extraordinary development of knowledge could eventually lead to an evolution not unlike the development of primates into human beings;

Therefore, in consideration of all the above, the participants of the First World Congress of Transdisciplinarity (Convento da Arrábida, Portugal, November 2–7, 1994) have adopted the present Charter, which comprises the fundamental principles of the community of transdisciplinary researchers, and constitutes a personal moral commitment, without any legal or institutional constraint, on the part of everyone who signs this *Charter.*

Article 1:

Any attempt to reduce the human being by formally defining what a human being is and subjecting the human being to reductive analyses within a framework of formal structures, no matter what they are, is incompatible with the transdisciplinary vision.

Article 2:

The recognition of the existence of different levels of reality governed by different types of logic is inherent in the transdisciplinary attitude. Any attempt to reduce reality to a single level governed by a single form of logic does not lie within the scope of transdisciplinarity.

Article 3:

Transdisciplinarity complements disciplinary approaches. It occasions the emergence of new data and new interactions from out of the encounter between disciplines. It offers us a new vision of nature and reality. Transdisciplinarity does not strive for mastery of several disciplines but aims to open all disciplines to that which they share and to that which lies beyond them.

Article 4:

The keystone of transdisciplinarity is the semantic and practical unification of the meanings that *traverse* and *lie beyond* different disciplines. It presupposes an open-minded rationality by re-examining the concepts of "definition" and "objectivity." An excess of formalism, rigidity of definitions and a claim to total objectivity, entailing the exclusion of the subject, can only have a life-negating effect.

Article 5:

The transdisciplinary vision is resolutely open insofar as it goes beyond the field of the exact sciences and demands their dialogue and their reconciliation with the humanities and the social sciences, as well as with art, literature, poetry and spiritual experience.

Article 6:

In comparison with interdisciplinarity and multidisciplinarity, transdisciplinarity is multireferential and multidimensional. While taking account of the various approaches to time and history, transdisciplinarity does not exclude a transhistorical horizon.

Article 7:

Transdisciplinarity constitutes neither a new religion, nor a new philosophy, nor a new metaphysics, nor a science of sciences.

Article 8:

The dignity of the human being is of both planetary and cosmic dimensions. The appearance of human beings on Earth is one of the stages in the history of the Universe. The recognition of the Earth as our home is one of the imperatives of transdisciplinarity. Every human being is entitled to a nationality, but as an inhabitant of the Earth is also a transnational being. The acknowledgement by international law of this twofold belonging, to a nation and to the Earth, is one of the goals of transdisciplinary research.

Article 9:

Transdisciplinarity leads to an open attitude towards myths and religions, and also towards those who respect them in a transdisciplinary spirit.

Article 10:

No single culture is privileged over any other culture. The transdisciplinary approach is inherently transcultural.

Article 11:

Authentic education cannot value abstraction over other forms of knowledge. It must teach contextual, concrete and global approaches. Transdisciplinary education revalues the role of intuition, imagination, sensibility and the body in the transmission of knowledge.

Article 12:

The development of a transdisciplinary economy is based on the postulate that the economy must serve the human being and not the reverse.

Article 13:

The transdisciplinary ethic rejects any attitude that refuses dialogue and discussion, regardless of whether the origin of this attitude is ideological, scientistic, religious, economic, political or philosophical. Shared knowledge should lead to a shared understanding based on an absolute *respect* for the collective and individual Otherness united by our common life on one and the same Earth.

Article 14:

Rigor, *opening*, and *tolerance* are the fundamental characteristics of the transdisciplinary attitude and vision. *Rigor* in argument, taking into account all existing data, is the best defense against possible distortions. *Opening* involves an acceptance of the unknown, the unexpected and the unpredictable. *Tolerance* implies acknowledging the right to ideas and truths opposed to our own.

Final Article:

The present *Charter of Transdisciplinarity* was adopted by the participants of the first World Congress of Transdisciplinarity, with no claim to any authority other than that of their own work and activity.

In accordance with procedures to be agreed upon by transdisciplinary-minded persons of all countries, this *Charter* is open to the signature of anyone who is interested in promoting progressive national, international and transnational measures to ensure the application of these Articles in everyday life.

Convento da Arrábida, November 6, 1994

Editorial Committee
Lima de Freitas, Edgar Morin
and Basarab Nicolescu

Appendix II

Table 1. Comparison between disciplinary knowledge (DK) and transdisciplinary knowledge (TK)

Disciplinary Knowledge	Transdisciplinary Knowledge
IN VITRO	IN VIVO
external world—Object	correspondence between the external world (Object) and the internal world (Subject)
knowing	understanding
analytic intelligence	a new type of intelligence a balance between intellect, feelings, and the body
orientation towards power and possession	orientation towards astonishment and sharing
binary logic	logic of the included middle
exclusion of values	inclusion of values

Appendix III

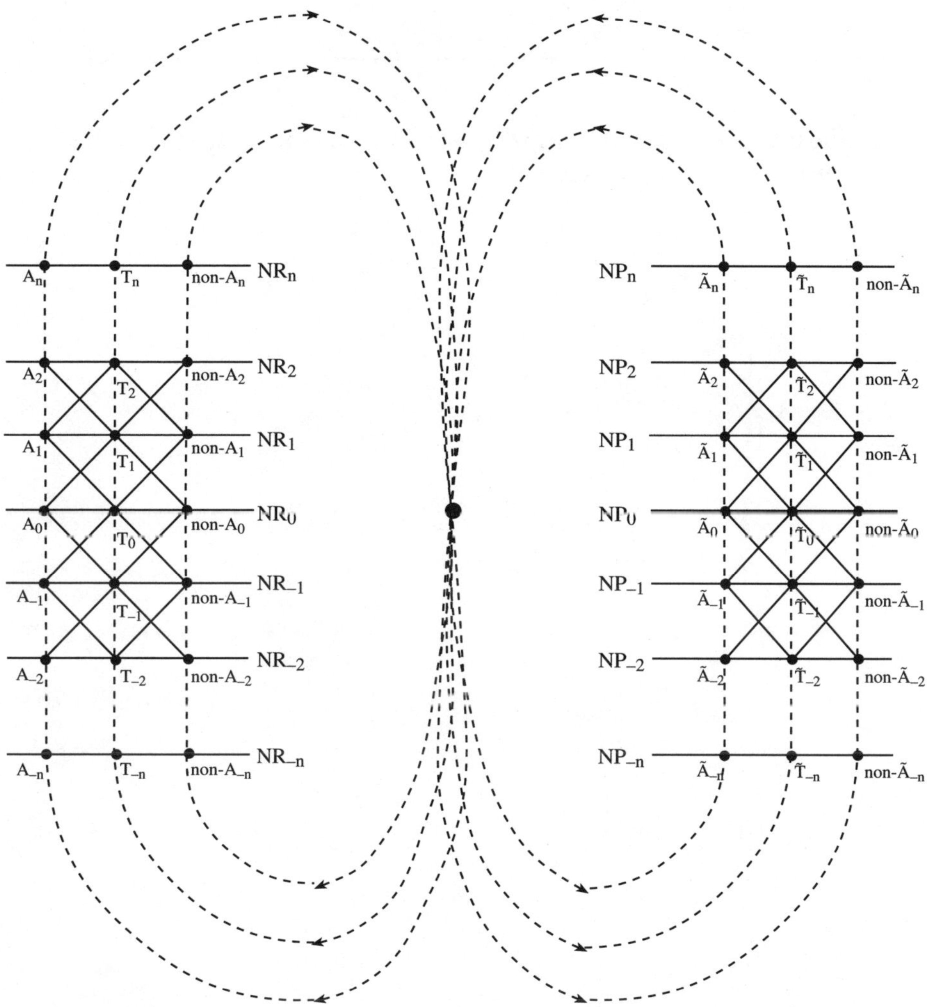

Figure 1. The transdisciplinary Object, the transdisciplinary Subject, and the Interaction term.

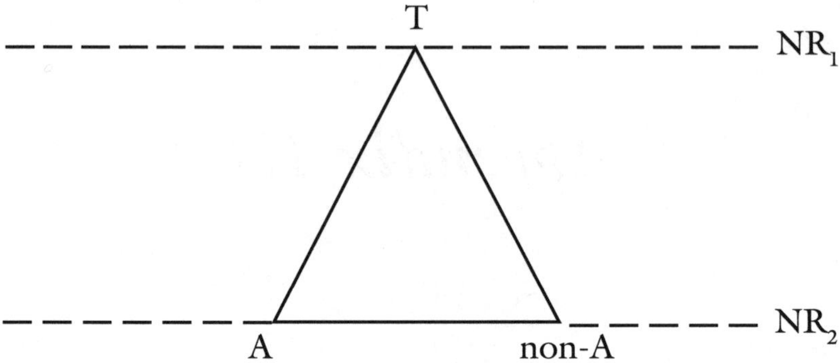

Figure 2. Symbolic representation of the action of the logic of the included middle.

Index